Praise for *40p Each or 2 f*

In *Forty Pence Each or Two for a Pound* Danielle Bartram ⎟ twice as often as she does the word 'hard', which gives us a strong indication of the application and usability of the book's vast array of mathematical content. From cover to cover, there are practical ideas and effective assessment and classroom-management strategies specially designed to engage learners and increase whole-school participation in maths.

I wish Danielle had been my maths teacher – things may just have turned out differently!

Ross Morrison McGill @TeacherToolkit

Dinner is over. The bill arrives. All heads turn expectantly to the maths teacher in the group and the inevitable question is posed: 'Well, how much should we each pay?'

Let's be honest, this transaction should be possible without a trained mathematician. The scenario is often mirrored in schools too, with numeracy skills seen as the sole responsibility of maths teachers. In this helpful book, however, Danielle offers her top tips and provides some really practical, flexible lesson ideas to nurture numeracy in classrooms across the school, not just in the maths department.

Chris Smith @aap03102, maths teacher, Grange Academy,
member of the *TES* maths panel and the Scottish Mathematical Council

The wonderful Danielle Bartram makes numerical literacy accessible to all in this masterclass which explains numeracy in context and gives it the boost needed to compete with its literacy cousin. Containing engaging topics, fun activities and contextual learning, *Forty Pence Each or Two for a Pound* will not only excite mathematicians, but also anyone who wants to promote a love of numeracy.

This could be a missing Teachers' Standard.

Deborah Barakat @mrsmathia,
Assistant Principal and Initial Teacher Training Lead, Excelsior Academy

'Numeracy across the curriculum' is a phrase that I have never truly understood – something that is even more worrying given that I have been a school numeracy coordinator in the past. Does it just mean doing times tables in French and scattergraphs in geography? Now I know that it doesn't.

With 31 activity types discussed, complete with teacher notes and variations, numeracy will no longer feel like an inconvenient bolt-on for non-maths teachers. This is an important book for those who have responsibility for sharing numeracy across their school, for those who aspire to do so and for the ever-growing numbers of non-maths specialist teachers wanting practical strategies to help mathematically reluctant students develop a love for the subject. I just wish this book had been around many years ago.

**Craig Barton, maths teacher,
maths adviser, *TES*, creator, mrbartonmaths.com**

Danielle starts *Forty Pence Each or Two for a Pound* with an honest and open introduction that will resonate with a lot of readers, and her keen enthusiasm for maths shines throughout the book as you can actually hear her voice talking you through the processes. Danielle reinforces the key message that every teacher is a teacher of maths, and that maths is intrinsic in what we do. Numeracy can easily be enhanced through the use of her suggestions.

Forty Pence Each or Two for a Pound provides a unique and coherent structure in the form of a numeracy chain which tackles six key themes. Using clear instruction, Danielle clarifies the multiple value of each resource as she guides you through a variety of tried and tested classroom approaches.

Invaluable reading for anyone promoting numeracy across a school curriculum.

Julia Smith, author and freelance maths teacher-trainer

Danielle Bartram manages to demystify mathematics by bringing the subject to life in a highly practical book that contains a plethora of ideas and activities. *Forty Pence Each or Two for a Pound* will build confidence in teachers looking to incorporate numeracy into their lessons and will be particularly useful as a tool to support the development of policy across the school.

David Bartram, Director, Prescient Education Ltd

Forty Pence Each or Two for a Pound is the essential guide that anyone with responsibility for numeracy in their school will want on their bookshelf, and on their mobile e-readers too.

Packed with practical ideas for all, and written by someone who has transformed numeracy in her school and knows how to make numeracy count, *Forty Pence Each or Two for a Pound* is a call to action for anyone who would ever say 'I don't do maths'.

Mark Anderson @ICTEvangelist, Director, ICT Evangelist, educator, speaker, consultant, award-winning blogger and author

Being numerate as well as literate underpins our daily lives, and it can be the key to unlocking many doors. In *Forty Pence Each or Two for a Pound* Danielle puts maths into an everyday context and enables teachers across all key stages to reach the disengaged student, the frightened-of-maths student, the 'I can't do it' student, and those who have been told by their parents, 'Don't worry, maths was never my strong point either.'

Forwards is backwards sometimes, and in this book Danielle enables teachers to go back to make sure the foundations of mathematical learning are secure before moving onto the next step. As mathematical learning starts in the early years foundation stage, *Forty Pence Each or Two for a Pound* provides the key messages that all teachers of maths and numeracy should know in order to support them in enthusing, enabling and empowering students with a love of mathematical learning. We may not all be teachers of mathematics, but we are all teachers of numeracy.

Students often ask, 'Miss, why do I need maths?' *Forty Pence Each or Two for a Pound* answers this question by highlighting the fundamental basics of maths and by showing us why we all need to ensure our students are savvy and well equipped so that they can live their lives (and avoid being ripped off) with a secure foundation in mathematical application and understanding.

Rachel Orr, education consultant, teacher, tutor and author

40p each | or 2 for £1

Making maths memorable, accessible and relevant

Danielle Bartram

Crown House Publishing Limited

www.crownhouse.co.uk

Published by

Crown House Publishing
Crown Buildings, Bancyfelin, Carmarthen, Wales, SA33 5ND, UK
www.crownhouse.co.uk

and

Crown House Publishing Company LLC
PO Box 2223, Williston, VT 05495, USA
www.crownhousepublishing.com

© Danielle Bartram, 2017

The right of Danielle Bartram to be identified as the author of this work has been asserted by her in accordance with the Copyright, Designs and Patents Act 1988.

First published 2017.

All rights reserved. Except as permitted under current legislation no part of this work may be photocopied, stored in a retrieval system, published, performed in public, adapted, broadcast, transmitted, recorded or reproduced in any form or by any means, without the prior permission of the copyright owners. Enquiries should be addressed to Crown House Publishing.

Crown House Publishing has no responsibility for the persistence or accuracy of URLs for external or third-party websites referred to in this publication, and does not guarantee that any content on such websites is, or will remain, accurate or appropriate.

Image page 10 © raven – fotolia.com, image page 40 © 3D Sparrow – fotolia.com, image page 43 © Maksym Yemelyanov – fotolia.com, image page 45 © jirikaderabek – fotolia.com, images page 46 © valdis torms and asfianasir – fotolia.com, images page 48 © toonbilge, eshana_blue and ilyabolotov – fotolia.com, image page 50 © Gianluca D.Muscelli – fotolia.com, images page 52 © koya979 and jules – fotolia.com, image page 73 © koya979 – fotolia.com, image page 86 © Alfonsodetomas – fotolia.com, images page 118 © T shooter – fotolia.com, image page 171 © Gribanessa – fotolia.com.

British Library of Cataloguing-in-Publication Data

A catalogue entry for this book is available from the British Library.

Print ISBN 978-178583012-9
Mobi ISBN 978-178583293-2
ePub ISBN 978-178583294-9
ePDF ISBN 978-178583295-6

LCCN 2017957031

Printed and bound in the UK by TJ International, Padstow, Cornwall

Acknowledgements

I would like to thank the teachers from far and wide, Twitter and beyond, who have supported me in my journey. Miss B's Resources started as a place where I backed up my electronic resources to now being a site used by hundreds of thousands of teachers worldwide. There are far too many people to mention individually who have supported me on this journey. However, a particular thanks must go to Lesley Ann McDermott (@LA_McDermott) and Barry Dunn (@SeahamRE), who both gave me the confidence to write a book and helped me realise there is value in what I have to share. I would also like to say thank you to Fiona Ritson (@FKRitson) who encouraged me to make a difference at the start of my journey as a numeracy coordinator. Finally, a special thank you must go to my editor, Peter Young, who helped me to transform my vivid imagination of resources into this book.

The amazing staff and students at my school need to be thanked for continually supporting me with my ventures and ideas.

Andrea Ayre, Michelle Dunning and Maria Gardner, what can I say? You are my rocks and my maths family. It is a true honour to work in a faculty of friends. Nothing is ever too much for you lovely ladies. You have kept me sane and inspire me every day. You are my very own cheerleading team.

Andrea Crawshaw, Michael Laidler, Sarah Ledger and Jon Tait, you took a chance on me for the job of lead practitioner and have always believed in me. No matter how crazy an idea is, you always bend over backwards to help and support me. My love of learning is kept alive through working with inspiring leaders such as yourselves.

Finally, thanks really does need to go to my family and friends who have been patient and understanding throughout the process of writing this book.

Contents

Acknowledgements .... *i*

Introduction 1

Chapter 1. What is Numeracy? 3

Chapter 2. Numeracy4All Chain 19

Chapter 3. Breaking Up the Journey 23

Chapter 4. Numeracy Links 39

Chapter 5. Subject Knowledge 57

Chapter 6. The 31 Prime Resources and Ideas 63

Literacy: **67**

    1.   Scrabblecross 68

    2.   Writing Weigh-In 71

    3.   Mathematical Language in Extended Writing 75

    4.   Talk Time 77

    5.   True Value 79

Exploration: **83**

    6.   Weight of the World 84

    7.   Impact Line 88

    8.   Headline Figure 92

    9.   Code Breakers 95

Engagement: **99**

    10. Twisted Figures 100

    11. Bargain Words 103

    12. Put the Fire Out 107

    13. Fuel Fill Race 110

14. Shopping Spree 113

15. Battle Words 115

16. Netting Questions 119

Marking and Reflection: **123**

17. Thermometer of Understanding 124

18. The Real Value 128

19. Maths Marking 131

20. Graph It 133

21. Results 136

Organisation and Presentation: **139**

22. Up to Date 140

23. Diagram Scales 143

24. Venn Diagrams 146

25. Two-Way Tables 150

26. Going with the Flow 152

Classroom Management: **159**

27. Measure of Success 160

28. Timers 165

29. Weight the Task 168

30. Groupers 170

31. Mix Up 173

Chapter 7. Enthusiasm 175

List of Resources *179*

References *181*

Introduction

Numeracy is a topic many people shy away from. It could be said to be a Marmite subject – you either love it or hate it – and for the majority of the population it is the latter. Yet there really are people who love maths. It's why I ended up becoming a teacher. And now I choose to work with those who don't share my passion because I appreciate how maths underpins so much of everyday life.

There is no doubt that maths is a subject many people struggle with or even fear. At secondary school I also found mathematics difficult. I had to work hard and grapple with the subject before it made sense to me. Therefore, I am well-placed to understand the difficulties that many students have in coming to terms with the maths syllabus.

I'm always open and honest about my reasons for becoming a maths teacher and why I do what I do. Maths was not a strength of mine, nor was any other subject. I had to apply myself to keep up with the elite. While studying for my GCSE in maths, my teacher was unfortunately suffering from mental health issues and as a class we went through many supply teachers, and that didn't help.

Since becoming a teacher I have worked tirelessly to make maths approachable for all students. I obtained a degree in mathematics from Lancaster University and did my PGCE at Durham University. Since qualifying, I have created Miss B's Resources (www.missbsresources.com) and regularly develop and share teaching and learning resources and ideas. I put particular emphasis on numeracy across the curriculum and especially in mathematics.

Because maths is intrinsically linked with many things throughout everyday life, I promised myself I would help to support as many people as I could who aren't natural mathematicians to have success in the subject. During my time at Lancaster University, I created a maths society which gave maths students access to more contact and support time with lecturers and PhD students on homework and study tasks.

Maths is about more than just numbers, letters and symbols on a page. It's about logic skills, making connections and solving problems. It's about gaining a deeper understanding of what things mean, how things work and why some decisions are better than others. For many people, it's a way of helping to explain some of the deeper patterns of the world.

Throughout its history, mathematics has often developed from people's obsessions. Many a mathematician has become passionate about a particular strand or topic of mathematics, and this has frequently produced breakthroughs in understanding. However, my obsession is much simpler: my aim is to make basic maths skills accessible and intrinsic to all students. To do this, they need to make connections with the subject content outside of the maths classroom, and they need opportunities to practise the basic skills on a regular basis.

For some students, numeracy across the curriculum seems to be the long-lost cousin to literacy across the curriculum. Both have a deep level of importance for everyday living. In practice, these two can be developed hand in hand. Having helped to roll out numeracy across the curriculum schemes in several schools, I felt it was time to bring together some of the practical tips and common-sense strategies that I have been using to help teachers with a desire to incorporate numeracy into their lessons or a policy across a whole school. There are ideas in this book suitable both for school numeracy coordinators and for primary and secondary teachers.

Chapter 1

What is Numeracy?

On occasion, I've seen grown adults reduced to a quivering heap by maths. Although this might be an extreme reaction by the unfortunate few, the word 'maths' seems to have become almost a swear word for some people. Many maths phobics draw on negative memories of school maths lessons and baffling concepts such as algebra or logarithm tables. Recollections like these are often where lifelong problems with maths begin.

However, there is a real issue in some educational establishments which treat numeracy as the estranged cousin to literacy. The thinking goes that so much of our lives rely on computers and algorithms to do things for us, that a basic knowledge of maths is no longer necessary, and, as such, maths gets placed on the back burner. This has disastrous consequences. The number of people who are innumerate is growing steadily and this needs to be tackled in the early years at school. With qualifications in subjects such as geography and science acquiring heavier maths core skills content, this reason alone should motivate us to embed links to numeracy across the full curriculum.

We live in a culture in which people go to extreme lengths to hide the fact that they can't read or write. Bizarrely, however, it is deemed socially acceptable to say, 'I can't do maths' or 'I'm no good at maths', and then do nothing about it. It's easy to forget the damaging effects that repeating those simple words to yourself can have on the growing mind; it then becomes a self-fulfilling prophecy. Parents, teachers, celebrities, even movies and advertising campaigns, may often deliberately or unconsciously promote people's perception that it is okay to be rubbish at maths.[1]

So, how can we expect students and the future leaders of tomorrow to hold maths and numeracy in high regard when society is continually telling them that being bad at maths is acceptable? According to a YouGov poll commissioned by the charity National Numeracy, regardless of the participant's level of the

---

1   See R. Garner, Shame celebrities who boast about poor maths, says numeracy charity, *The Independent* (15th September 2014). Available at: http://www.independent.co.uk/news/education/education-news/ shame-celebrities-who-boast-about-poor-maths-says-numeracy-charity-9734152.html.

subject, 80% either strongly agreed or agreed with the statement, 'I would feel embarrassed to tell someone I was no good at reading and writing.' However, only 56% of people either strongly agreed or agreed with the statement, 'I would feel embarrassed to tell someone I was no good with numbers and maths.' However, the issue is even more prevalent with females: 82% saying they would feel embarrassed to tell someone, 'I was no good at reading and writing' and only 53% saying they would feel embarrassed to tell someone, 'I was no good with numbers and maths.'[2] This is a 29% difference with the female sample compared to the male sample, where the difference was only 18%.

In failing to change this culture of ignorance, we are handicapping ourselves for the rest of our lives. No matter how much we might loathe mathematics, we need to acknowledge that we all use it on a daily basis. We are surrounded by mathematical concepts all day, every day. Therefore, we need to change the passive acceptance of failure and find every way possible to support students to overcome their maths hang-ups.

It is the job of the teacher to be not only numerate themselves, but also to recognise and respond to any numeracy weaknesses in their students. It is only in this way that progress will be made, and this has to happen in *all* areas of the curriculum as well as in everyday life out of school. As a result, students become more competent and are more likely to buy into their learning.

The first part of this process is to make it acceptable to admit to poor numeracy skills, as that is the basis for doing something about it. The acceptance of poor numeracy skills is different from the proud declaration of not being able to do maths which can lead to students who won't try and who seem not to care. There should be no shame – or glory – attached to a student's admission. Instead, it should be seen as part of learning to do better, demonstrating a growth mindset and the foundation of future success in life.

---

2   See http://cdn.yougov.com/cumulus_uploads/document/tm2q3p27f6/Results-for-National-Numeracy-Numeracy-10032015.pdf.

Why maths matters

Although it is not possible to be definite about how the working environment will change in the future, we can be certain that it will change, and that some kinds of work will be taken over by automation and artificial intelligence, reducing the amount of skilled and semi-skilled work that we became familiar with in the 20th century.

In other areas jobs will rise. The National Careers Service predicts that 'Employment in the trade, accommodation and transport industries is expected to increase by 400,000 jobs by 2020. Much of this growth will be in distribution, retail, hotels and restaurants.'[3] This means that in the future there will be an increasing need for people to work in catering services; social services; sport, leisure and tourism; transport; storage, dispatch and delivery; and retail sales. All of these require numerate workers, capable of measurement, timetabling, costing, statistical analysis, networking, accounting and forecasting. In addition, many people will be expected to be fluent in computer technology and able to write their own algorithmic software programs.

If the UK is to continue to develop economically, then today's students need to understand the relevance of numeracy, and that unless they are able to participate in future developments in technology, they will find it difficult to get work. Research carried out by Pro Bono Economics, drawing on a number of factors, estimates that low numeracy levels cost the UK £20.2 billion (1.3% of GDP) each year.[4] A massive 68% of employers are concerned about their employees' ability to understand if the figures presented to them make sense – what is commonly called a 'sense check'.[5]

So, maths really does matter in the current climate, and for the future if the economy is going to grow and adapt to changes in technology and working practices. As educators, we need to make sure that companies have faith that

---

3   See UK Commission for Employment and Skills, *Working Futures 2010–2020. Evidence Report 41* (August 2012). Available at: http://webarchive.nationalarchives.gov.uk/20140108090250/http://www.ukces.org. uk/assets/ukces/docs/publications/evidence-report-41-working-futures-2010-2020.pdf.

4   Pro Bono Economics, *Cost of Outcomes Associated with Low Levels of Adult Numeracy in the UK. Pro Bono Economics Report for National Numeracy* (2014). Available at: http://www.probonoeconomics.com/sites/ default/files/files/PBE%20National%20Numeracy%20costs%20report%2011Mar.pdf, p. 4.

5   National Numeracy, *Numeracy Review* (Lewes: National Numeracy, 2015). Available at: https://www. nationalnumeracy.org.uk/sites/default/files/numeracy_review_overview_v2.pdf, p. 2.

our young people have a sufficiently high level of mathematical ability, so they remain highly employable in both the national and global market. Improving numeracy within schools is a fundamental way of developing core skills.

Numeracy is the stepping stone that allows students to access mathematics. It is, if you wish, a subset of mathematics. Numeracy is the basic ability to recognise and apply simple mathematical concepts to solving problems in everyday life. It includes basic skills such as addition and multiplication, which enable us to handle common functional maths topics such as weighing ingredients and telling the time. In contrast, the more complex domains of mathematics – such as algebra, trigonometry, calculus or topology – are a minority interest and most people leave them behind in the classroom.

However, mathematical reasoning provides a way to develop the mind and train the thought processes needed for problem solving. It is these basic skills that transfer to real-world problems. Some of the topics learned in the maths classroom may seem irrelevant, but it is important that we develop the analytical and logical thinking skills which will support future learning and comprehension.

Figure 1.1 demonstrates how numeracy underpins the resources and principles in this book. In short, without the foundation of numeracy, students will be unable to understand or fully access the world around them.

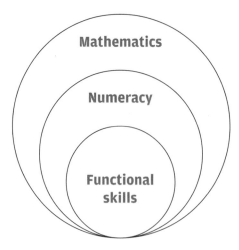

Figure 1.1

It's not okay to be innumerate

There is a social stigma attached to calling someone illiterate, but this does not apply to being called innumerate. This could be because the need for literacy is an essential everyday requirement. If you can't read you will struggle to perform many tasks that others find easy. It soon becomes obvious to others that you can't read because you can't understand written information – for example, that key aspect of modern life, the ability to send and receive texts. Literacy, unlike numeracy, is a visible skill that students can observe everywhere they go from a very early age. Conversely, many numerical tasks are performed automatically, such as on cash-registers in shops and online payment systems, and this seems to remove the requirement for mathematical competence.

Numeracy is often a more hidden skill that lurks in the depths of daily tasks. Young children can get by without a basic level of numeracy, but as they grow older the need for competence in numbers becomes an essential part of getting a job and in taking on the responsibilities of a householder. Numeracy is no longer a hidden skill.

Herein lies the basic problem. Students can't comprehend what is needed in the future if they don't know what to expect. Just telling them isn't good enough. They need a clear understanding of why they are being asked to do things, and for many hard-core students who refuse to admit a need for numeracy skills, seeing is believing. Unfortunately, many students only realise the purpose and need for basic numeracy skills too late.

It is therefore essential that, as teachers, we provide opportunities for students to understand why they are doing things and the benefits they will get from learning particular aspects of numeracy. Unlike GCSE mathematics, where students study many things that they will probably never use again afterwards, it is important to place the need for numeracy in context as an essential life skill *before* we begin to teach the specific techniques they need to acquire.

There are two types of numeracy issues with students. The first is with those who can't be bothered to learn the skills because they can't see the purpose of them. The second is a much more delicate situation. It concerns those who struggle with mathematics and who may even suffer from dyscalculia. These students may get a lot of support and encouragement, but unfortunately – given the way the education system is set up – this is often designed to help them keep up with the material that other students are studying. Because they are moving at a pace too fast for deep understanding, these interventions can lead to ever more superficial understanding, with many gaps and little retention.

The core principle of numeracy is for the foundations to be secure. It's the same as building a house in an earthquake zone: if the foundations are dug deep enough and the building constructed with the right materials, the house will be much less likely to collapse. In practical terms, the way to do this – instead of doing what the curriculum dictates, which is continually pushing forward those students who struggle – is to go *backwards*. This means playing with the games and activities in this book and using mathematical equipment to help embed and support the principles of teaching for depth carried out by the maths faculty. This will help to ensure that students understand the most basic number skills, with no tricks or gimmicks. When done correctly, students will have a greater rate of retention with mathematical topics.

I started this 'back to basics' technique with students in my second year of teaching. I have found that when it comes to teaching mathematics, schools often give the lowest ability Key Stage 4 groups to the newest teachers. This is certainly what happened to me. I took on students who were all predicted U's or G's, and I have to admit that I didn't succeed with all of them. However, by applying the principles above, I managed to help motivate one student to access enough content to gain a grade C and several others achieved D's and E's. There were no longer any U's, but some never got beyond F's and G's. I applied this same principle again the following year with a similar set with even more success. However, it took a lot of buy-in at this stage to improve a bottom set at the start of Year 11. But imagine if these principles were applied early on: how much more would the students be able to access? Would they learn to love maths instead of loathing it?

Another problem with a lot of maths teaching is that students are instructed about the mathematical techniques and operations without being provided with a clear understanding of the purpose or desired outcome – that is, what you want to do and what kind of answer is expected. In other words, the big picture is missing. Therefore, the teacher needs to explain *why* they are doing these things and what benefit they will get, rather than merely concentrating on the nitty-gritty.

As a maths teacher, one of the most common things I get asked is, 'Miss, why will I need this?' or, worse, I am told, 'I will never need this in real life, what's the point?' However, I've often found students to be extremely curious, and they are not being deliberately awkward in asking these questions. Since they have to learn and retain so much information on a daily basis, of course it is important that they prioritise what they are taking in. I have discovered that the best way to do this is to show students non-hypothetical examples of where they will use elements of mathematics in real life – I will cover many examples in this book. I am also honest that the likelihood of them using particular skills in life is slim, such as the sine rule or the cosine rule, but I do go on to explain the reasoning behind why it is still important to learn it and the significance of the thought processes it builds.

A key example of this understanding clicking into place for a student of mine was when he was completing a Duke of Edinburgh's Award scheme trek. This meant they would have to use a map and compass to find their way. As I knew that this student would be on the trek, I thought this would be a great time to tackle his stubbornness as a mathematics learner. So the week before I made sure I covered bearings and map-reading skills. On returning to school the following Monday, the student was in awe that his maths lesson had actually helped him with something tangible – in his words, 'It totally wasn't maths, Miss, but what we did helped us not get lost.' We then followed up this statement and had a brief class debate on how sometimes the things we least expect to be of use can actually help us the most.

There are many ways in which elements of a maths lesson can instantly relate to a student's everyday life experiences. Here the students get to see numeracy in action and can relate to the awe of the student described above.

Numeracy in action 1: shopping

Students need to be able to leave education with the ability to handle the tasks of everyday life. Some are unable to deal with simple monetary amounts; others struggle when they need to accurately scale a recipe up or down when cooking for more or fewer people than is stated in the recipe. These are everyday skills that students often don't link to mathematics.

Now, if they can't calculate monetary amounts, how do they know if they are being ripped off? For example, Figure 1.2 shows some genuine shop price labels. I've shown these to staff and students and asked them to spot what is wrong. Sometimes it takes a while for them to understand what the mistake is and how they are being misled.

Figure 1.2

These labels all show errors of one kind or another, but I wonder how many students would realise this and check the calculations. For example, the offer of 40p each or two for £1 means students would need to calculate a very simple sum to work out that it is actually cheaper to ignore the two-for-one offer.

How often do we miss these mistakes ourselves when shopping? Comparing the price per 100g or per 100 sheets of toilet paper is an essential skill, and enables us to get the best value for money rather than being deceived by pseudo offers. When you have a limited budget, every penny counts. Being able to calculate unit pricing is a vital numerical skill that students need to master. This comes down to their number sense and problem-solving skills. For example, they need to understand that 10.5p is more than 10.05p. I've purposefully used this example as it is a common mistake students make when attempting to put decimal numbers in ascending order.

Stores often try to direct customers to offers or multi-buy deals that they want to promote, but they also place a premium on those brands and products which earn brand loyalty from consumers. How many of us regularly calculate the benefits of each offer when a supermarket places rival offers from well-known brands on an end-of-aisle display?

Figure 1.3

Let's have a look at an actual choice in a supermarket (Figure 1.3). Which offer would you go for? How long would you spend evaluating the alternatives while standing in the aisle? Many people go for speed and purchase the 'Puppy' offer. It looks like a clear bargain: you gain the third pack for only £2.50. But is it really a bargain?

There are three different levels of analysis you can use for comparing these types of offer. The first technique for making a fair comparison is to calculate the price per roll. Many students will write and calculate the division the wrong way round. For example, to find the unit price of 'Silk' toilet roll they calculate 24 ÷ 7, instead of 7 ÷ 24, often because they believe you always divide by a smaller number. Other students may not even realise that they need to divide.

Panda toilet roll	**Puppy toilet roll (3 for £10 offer)**	**Silk toilet roll**
9 rolls costs £3.50	27 rolls costs £10	24 rolls costs £7
£3.50 ÷ 9 = 0.38**89**	£10 ÷ 27 = 0.37**04**	£7.00 ÷ 24 = 0.29**17**
1 roll costs **39p**	1 roll costs **37p**	1 roll costs **29p**

Note: All rounded to two decimal places.

From the calculations, 'Silk' is clearly the best offer by a long way, but do you get more or fewer sheets per roll (PPS or price per sheet)?

Panda toilet roll	**Puppy toilet roll (3 for £10 offer)**	**Silk toilet roll**
1 roll with 180 sheets costs **39p**	1 roll with 160 sheets costs **37p**	1 roll with 200 sheets costs **29p**
PPS = 39 ÷ 180	**PPS** = 37 ÷ 160	**PPS** = 29 ÷ 200
= **0.22p**	= **0.23p**	= **0.15p**

Note: All rounded to two decimal places.

This second level of analysis, for students who understand place value, provides an opportunity to compare products more accurately and in depth. The calculations show that 'Silk' is the best offer by a significant margin.

The comparison per roll shows that 'Puppy' seemed to be better value than 'Panda'. However, the more in-depth per sheet calculations show that the 'Puppy' offer isn't as good as the 'Panda' offer. Supermarkets often write these comparison figures in small print on the product labels.

When you are in a supermarket, how much time do you spend comparing products in terms of the *quantity* of the item you get, whether it is eight chocolate bars or nine toilet rolls? Do you work out how much product you are getting in precise detail, or do you simply assume that the products are like for like?

The third level of analysis that you could carry out when deciding on the best product is to compare the *quality* of the product. Both the 'Panda' and 'Puppy' brands are 2-ply whereas the 'Silk' brand is 3-ply – so the offer that many shoppers rejected was by far the best deal.

How many times do you miss out on the best deal due to brand loyalty or simple oversight? In times of austerity, many families cannot afford not to find the best offers simply because they are lacking basic maths skills and don't know how to calculate the appropriate sums, either in their head or with the calculator on their mobile phone.

Numeracy in action 2: utilities

Money matters involve more than just buying items in a shop. Another common functional maths GCSE question involves utility bills and meter readings. Examiners frequently report that students do badly on these. With increasing competition between gas and electricity suppliers, and frequent price fluctuations and offers, we need to be able to calculate which deals offer the best value.

Account Number
0112358 13 21

Electricity Bill

Meter number
Prime 73 1001001

Previous reading	Current reading	Unit price (pence)	Units used
6147	6529	7	?

Total:_____

Step 1: Calculate the amount of units used.
Current reading – Previous reading

$$6529 - 6147 = \textbf{382 units used}$$

Step 2: Calculate how much it is to pay for the electricity used.
Units used × price per unit

$$382 \times 7 = 2674\text{p}$$
Total = £26.74

The first major area of concern is that many people just don't understand their gas and electricity bills. If your meter reads 6529 then you need to know what this figure represents. As an absolute value it has little meaning. It's only when you calculate the difference between the present reading and the previous reading (one month or three months ago), that you can work out the amount of energy used during that period.

Many businesses use decimal amounts of pennies on charges (such as 6.9p per unit or 121.9p per litre), so students need to be aware that companies will round up because they can't charge you for the decimal amount. This also needs to be taken into account when you are buying fuel. Do you buy petrol at the closest filling station at a price of 119.9p per litre, or is it worth driving five miles to the next petrol station which is selling petrol at 119.7p? How much could you actually save? In this scenario, it may cost you more in petrol to get to the cheaper station than the money you would be saving.

Life skills, such as being able to go shopping within a budget and checking that your bills are correct, could easily be added to, or used to enhance, a school's guidance curriculum, as well as appearing in maths lessons. It is important that *all* students value these skills. Unfortunately, those lacking numeracy skills resist taking the necessary action to remedy this. Instead of listening to their maths teacher, they shut off, believing, 'I'm no good at maths' or 'I'm never going to need this.' Not true. These skills will play a vital role in their future. However, these students often will listen to teachers in other subjects using a less formal 'sneaky' maths session!

> Resources 5 (True Value), 11 (Bargain Words), 13 (Fuel Fill Race) and 14 (Shopping Spree) would be the go-to activities for this.

Numeracy in action 3: time

Many students and adults are unable to tell the time. And because they struggle with the basic concepts of telling the time, they miss out on events or always arrive late because they aren't aware of the time or they can't accurately estimate how long things will take.

A basic skill that all students need to be able to manage is their timetable and arriving to lessons on time. Lots of schools have moved to 'no bells' systems over the years, meaning that students need to be more aware of the time in order to arrive to lessons punctually. This becomes even more important when students move on to higher education where they need to be able to manage their 'free' periods. Another natural part of a school day for many students is being able to get to the bus stop on time in order to catch the bus to and from school. Students also need to be able to predict what time they will be home in order to keep their parents informed. However, many students don't work to precise timings and struggle to read a timetable, which means they are often unable to calculate journey times.

In addition, in an examination, students need to be able to understand the signs at the front of the hall, often written in digital clock form, about the start and finish times of the exam and calculate how much time they can spend on each question. They need to be able to apply this information to the analogue clock on display to calculate how much time they have left.

A good reality check in a secondary school environment is to ask your students to read the time and then calculate some different time intervals, such as, 'What will the time be after two 50-minute lessons?' Some students will find this impossible to work out. This will tell you where to direct your attention. All teachers need to help their students to become confident with these basic skills.

Resources 4 (Talk Time), 22 (Up to Date) and 28 (Timers) would be the go-to activities for this.

Numeracy in action 4: health

A common pitfall in students' mathematics skills is their inability to apply this knowledge in numerical or functional contexts which could have a serious impact on their lives or the lives of others. For example, if a child is ill and a parent or carer needs to give them medication, it's essential that they can work out the correct dose.

<table>
<tr><td>

These are fictional instructions, please don't use.

Junior Medicine

250g paracetamol per 5ml. Every pack contains a 5ml spoon. Suitable for children aged 5 or above.

Age	Dosage
Under 5 years	Consult doctor
Children 5–11 years	5–10ml every 6 hours
Children 12+	5–20ml every 6 hours

Leave 6 hours between doses and do not take more than 3 doses each day. If the symptoms have not cleared after 5 days please seek medical advice.

</td><td>

These are fictional instructions, please don't use.

Infant Medicine

125g paracetamol per 5ml. Every pack contains a 5ml spoon. Suitable for children aged 3 months plus.

Age	Dosage
Under 3 months	Consult doctor
Children 3–5 months	2.5ml every 6 hours
Children 5–12 months	2.5–5ml every 6 hours
Children over 12 months	2.5–20ml every 6 hours

Leave 6 hours between doses and do not take more than 3 doses each day. If the symptoms have not cleared after 5 days please seek medical advice.

</td></tr>
</table>

To demonstrate this, I often give teaching staff a set of pseudo medication instructions for paracetamol and ask them to answer the following questions:

- You have a 2-year-old child with a fever. Which medicine should you give?

- You have a 10-week-old child with a fever. Which medicine should you give?

- Your 5-year-old child has been given three doses of 5ml today. How much more medicine can they receive within the next 12 hours?

- You last gave your 3-year-old child a dose of medicine at 1 p.m. It is now 2 p.m. When can you give your child another dose?

I find this task hits home with a lot of staff about the true importance of being numerate. Many people do not realise that teaspoons vary in size – from less than 3ml to more than 7ml (i.e. in mathematical terms, 5±2ml) – and are therefore unlikely to deliver the correct dosage. In other words, an ordinary teaspoon should not be used for dispensing medicine. This is why a standard 5ml spoon is usually included with the medicine. Health considerations will be further impacted if the person is also not able to tell the time as they will be unable to calculate when to give the next dose.

> Resources 2 (Writing Weigh-In), 4 (Talk Time) and 28 (Timers) would be the go-to activities for this.

It is the job of all teachers – not just maths teachers – to ensure that students have plenty of opportunities to apply their mathematical skills outside of the maths classroom. As with any practice, the more dedicated time that students give to this, the more easily they will be able to adapt to new problems and situations. If our young people are to feel confident that they will be able to deal with life's challenges, then we need to breed a culture of importance around numeracy and maths, encompassing both the everyday skills and the specific skills needed for future prosperity.

So far I know:

- There is a negative culture towards numeracy which needs to be challenged.

- Poor numeracy skills can have a detrimental impact not only on an individual's quality of life and life chances, but also on the development of the economy.

- Numeracy is a subset of mathematical skills.

- Being numerically competent is an important skill in helping with the everyday tasks of running a household, including shopping, paying bills and taking medicine.

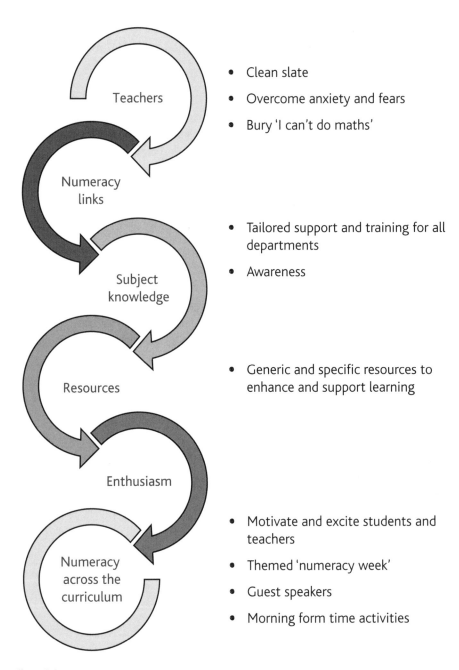

- Clean slate
- Overcome anxiety and fears
- Bury 'I can't do maths'

- Tailored support and training for all departments
- Awareness

- Generic and specific resources to enhance and support learning

- Motivate and excite students and teachers
- Themed 'numeracy week'
- Guest speakers
- Morning form time activities

Teachers

Numeracy links

Subject knowledge

Resources

Enthusiasm

Numeracy across the curriculum

Figure 2.1

Chapter 2
Numeracy4All Chain

Something I learned very quickly in the process of developing numeracy across the school curriculum is to start small but aim big. There will be many hurdles on the way to implementing literacy and numeracy across the curriculum, so stay strong and be prepared for setbacks. With persistence, resilience and purpose, your colleagues will join you and together you will build a positive momentum. Whether you are a teacher within a department trying to find openings for numeracy in your subject lessons or a coordinator/leader trying to persuade staff to include numeracy in lessons, the most important point to get across is never to accept – or allow the students to use – the statement, 'I can't do maths.'

Numeracy4All is exactly what it says it is: a way of ensuring that numeracy is essential to the lives of everyone. The Numeracy4All chain is designed to help meet this challenge and become part of the solution. It places numeracy at the heart of the curriculum and school life, empowering both teachers and students. Numeracy4All discourages the negativity surrounding maths and numeracy, particularly in secondary schools. It focuses instead on discovering the excitement behind core mathematics and numeracy skills.

My original chain had five links – teachers, enthusiasm, subject knowledge, resources and numeracy across the curriculum. I later added a sixth link – numeracy links – to represent the visual and physical embedment of numeracy across the curriculum.

The numeracy links are easily accessible and user-friendly resources – the glue that binds the numeracy chain together. It means that all teachers have the same basic standards and processes that will support students' learning. I have worked closely with teachers from every department/faculty to develop these links, and I have tailored them so that they are useful in every subject area. The links cover six different areas:

1. Functional skills

2. Graphs and statistics

3. Problem solving

4. Shapes and measures

5. Formulae and equations

6. Number

These six categories enable staff to clearly identify what areas of maths and numeracy appear in schemes of work and also to make this information available to students in a particular lesson. They also provide students with a go-to crib sheet for help and support in the six domains.

Planning for numeracy

If you are going to deliver a numeracy programme for staff in any school, you will need to spend time preparing. This will involve several planning sessions with key personnel before presenting the material.

Make sure you do your research and trial ideas in several departments and with key members of staff. It may be useful to set up a working party for the numeracy coordinator to lead, including representatives from different subject areas. This will then allow ideas to flow and promote communication across the school as staff share and disseminate ideas and practices. Note that what has worked in one school may not work in your school, so it is important to adjust the ideas to suit your particular students, staff and school ethos. To help make numeracy across the curriculum a success, and to avoid problems later down the line, it is essential to listen to and adapt to each subject's needs. And because every subject is different, you will need to gather a range of ideas and opinions.

As part of the delivery of the school's numeracy across the curriculum plan to all staff, it is a good idea to have examples of success stories from trials which teachers in your school have already run. These will help to generate interest and excitement in the staffroom about what is to come. I found that as numeracy levels increased across the curriculum, the enthusiasm also continued to grow.

The timing of the launch with students requires a great deal of forethought and planning. Although you may be keen to get started, rushing in is not always the best route. There should always be initial staff training on the numeracy links before it is launched with the students, perhaps during a numeracy week.

When embedding the numeracy links with staff, I work closely with departments on their individual needs and concerns. Where appropriate, I also suggest a few resource ideas to use in the lead-up to the launch of numeracy week. This gives staff and students a chance to get used to the idea and grow in confidence.

Once staff have had time to explore the distinction between numeracy and maths, they tend to feel more self-assured in approaching numeracy across their particular part of the curriculum. This process tends to bring to light several realisations, the most significant being the recognition that maths topics exist across the curriculum and that numeracy concepts are a vital component.

When done well, the Numeracy4All chain has enabled schools to create an interleaved and holistic curriculum which encourages and excites both staff and students. By breaking up the journey into manageable steps, it is possible for schools to instil the principles of the numeracy chain and links into whole-school and classroom practice.

So far I know:

- Numeracy4All ensures numeracy is essential to the lives of everyone and accessible to all.

- To lead by example in banning the statement, 'I can't do maths.'

- To make it clear that every teacher needs to be a teacher of numeracy.

- To break the challenge down into the numeracy chain and links when introducing numeracy across the curriculum.

Chapter 3
Breaking Up the Journey

Tackling teachers

Education policies change frequently and new initiatives arrive every year. It's not surprising that both teachers and students become jaded with these constant disruptions and revisions. Therefore, it's essential that you overcome any cynicism or resistance to promoting numeracy and demonstrate your own enthusiasm in order to entice, excite and enthuse staff and students about maths.

A proportion of teachers may have a maths phobia, often stemming from their early negative experiences of mathematics. Even the mere mention of anything related to numeracy or maths can create a strong fearful reaction. If teachers do have a negative attitude to maths, then it's very easy for them to unwittingly communicate this when they are talking generally or when responding to students. And if, 'I was never any good at maths' slips out, it can help to perpetuate the 'I can't do maths' culture. However, there is a difference between a fear of maths and being innumerate. These should not be confused.

As we have seen, maths is more widespread than many people realise, and often so deeply engrained in our daily practice that we don't think about it. For example, teachers continually use and interpret data to support arguments and discussions, as well as generate their own data on performance measures. They also have to plan, organise and cost trips. This is maths in practice.

Although many people in the profession have a fear of getting something wrong and embarrassing themselves, particularly in front of colleagues or students, the majority of teachers can perform basic numeracy and more complex mathematical reasoning tasks. We know this because to get a place on a teacher training programme applicants have to complete the QTS numeracy test and have a GCSE in mathematics. However, taking maths techniques into a classroom environment, and having the confidence to share and deliver these skills, can prove troublesome.

Half the battle in promoting numeracy can often be in overcoming the school's previous numeracy vision. I have worked with several schools whose old numeracy policy was a list of 'how to do maths' topics – and I do mean maths topics here, such as solving algebraic equations. One school's plan initially involved having a teaching and learning session simply for the coordinator to read the material out loud to the staff. This was based on the delusional expectation that all staff would then know how to cover all of the topics. On this occasion, the school's senior leaders stopped any further sessions as they realised this wasn't the way they wanted to go. However, the damage had been done, and the next steps forward in developing numeracy across the curriculum needed to start with winning back the trust of staff.

I believe that the education system overall has moved on from the misconception of maths and numeracy being the same thing, and towards an understanding that all teachers need to incorporate numeracy concepts into their lessons. However, simple acknowledgement is one thing; knowing how to proceed is a somewhat different problem.

To regain teachers' trust and enthusiasm for numeracy, staff need to be willing to wipe the slate clean and view numeracy through a fresh pair of eyes. It's important that staff do not dismiss the changes to numeracy across the curriculum policies without giving them a fair trial. It takes time to bed down any new system if it is going to stand a chance of success.

Teachers need to be involved in the change process from the get-go. Both schools and individual teachers must recognise what they need to do differently. Sometimes the old model can be adapted, but it's probably more the case that some new thinking is required for the change to happen. For a start, teachers need to think about how numeracy can be a key element in topics across the curriculum, not just in their own specialisation. Take the mathematics department, for example: one would hope that they would already be incorporating other subjects into mathematics lessons. However, the English department may only use numeracy when handling student data.

The importance of an audit

What I'm about to state might be seen as unwelcome in an already overstretched teacher workload, but consider carrying out a numeracy audit to find out the starting points of each department.

Please put the pitchforks away! I am aware that the idea of a numeracy audit will need to be sold to the staff, but there is a very clear purpose in doing this. The holistic view is that the numeracy coordinator, maths faculty and individual departments need to be able to develop their schemes of work to support each other. For example, in my current school, Year 9 look at standard form (using powers of 10 to express how big or small a number is) and the planets early on in the year in science lessons. However, in our old maths department scheme of work, some students wouldn't see standard form until the end of Year 9 or in Year 10. The old scheme of work had to change. Given how important numeracy is across the curriculum, wherever possible the maths department should consider discussing and possibly reordering the sequence of topics with other departments in mind.

The structure, format and length of the audit is extremely important in order to gain an accurate picture of numeracy, without overloading staff. This should be completed collaboratively as a department rather than individually; the intended aim is to give teachers a chance to reflect on and share their own practice, as well as listen to others. Hopefully, such meetings should initiate many great conversations and discussions (e.g. if a student writes today's date in their book, is this numeracy?).

The data from these audits on numeracy activities should be sought and collected from as many faculties in the school as possible. Where appropriate, this should include service provision for those students with extra educational needs because you need to build up a full picture of numeracy across the curriculum. The specific data you need to collect is whether particular maths concepts are used, and if so, in which subject areas, with which year groups and when.

This could take the form of a simple closed-question yes/no questionnaire for each department. In this case, I would advise that you then categorise the data you have collected into the six numeracy link categories in order to standardise the thinking across the whole school.

In this way, you will be establishing a baseline, so in the future you will be able to get a measure of progress with these different aspects of numeracy.

When collecting information, try to use the core skills that appear across the curriculum as these will allow for a later comparison – for example, adding and subtracting numbers are present in the objectives of many activities across the

Difficulty	Objective	Year 7		Year 8		Year 9		Year 10		Year 11	
		Y/N	HT	Y/N	HT	Y/N	HT	Y/N	HT	Y/N	HT
Bronze											
Silver											
Gold											

The list provided isn't exhaustive and may have missed skills you incorporate within your scheme. You may also have new skills that are a requirement within the new specifications. If this is the case, please list them below within the relevant year group. Please remember to provide the half term this skill is covered in when possible.

Additional comments

Year 7	
Year 8	
Year 9	
Year 10	
Year 11	

curriculum. It is also important to provide opportunities for staff to write additional comments. Do not overload the audit with objective statements that only apply to a minority of departments.

What follows is an example of the numeracy audit I have sent out to departments in the school I have worked with. The accompanying text is therefore directed at teachers/faculties.

Numeracy audit

The school wishes to perform a numeracy audit across the curriculum in order to develop a package to enhance learning in the classroom. As part of this programme, we want to be able to support all faculties with the numeracy and maths skills already incorporated in schemes of work, as well as gaining a picture of numeracy across the curriculum. Within the mathematics faculty, we are always trying to improve our new schemes of work and we feel this is the perfect time to carry out a numeracy audit.

How to complete:

Step 1: Please complete an audit relating to the six tables that follow for each year group. To do this, please place a Y in the yes/no column next to the objective if you use this skill in your scheme of work. If you are aware which half term (HT) this appears in, please write number 1 to 6 here: (1) Autumn 1, (2) Autumn 2, (3) Spring 1, (4) Spring 2, (5) Summer 1, (6) Summer 2 (please list all if several apply). This will enable us to cross-reference with the maths faculty scheme of work.

Step 2: Please complete the additional comments box with a brief description of when and how you use these skills.

Note: The learning objectives listed aren't exhaustive and may have missed skills you incorporate within your scheme. You may also have new skills that are a requirement within the new specifications. If this is the case, please list them in the bottom half of the tables within the relevant year group. Please remember to provide the half term this skill is covered in when possible.

Problem solving

Difficulty	Objective	Year 7		Year 8		Year 9		Year 10		Year 11	
		Y/N	HT	Y/N	HT	Y/N	HT	Y/N	HT	Y/N	HT
Bronze	Collect data										
Bronze	Create surveys and questionnaires										
Silver	Represent information graphically										
Gold	Consider sampling										
	Additional comments										
	Year 7										
	Year 8										
	Year 9										
	Year 10										
	Year 11										

Graphs and statistics

Difficulty	Objective	Year 7		Year 8		Year 9		Year 10		Year 11	
		Y/N	HT	Y/N	HT	Y/N	HT	Y/N	HT	Y/N	HT
Bronze	Use flow charts to plan processes										
Silver	Tackle open problems										
Silver	Present work and findings in a logical order with evidence										
Gold	Consider sampling										
Additional comments											
Year 7											
Year 8											
Year 9											
Year 10											
Year 11											

Formulae and equations

Difficulty	Objective	Year 7		Year 8		Year 9		Year 10		Year 11	
		Y/N	HT	Y/N	HT	Y/N	HT	Y/N	HT	Y/N	HT
Bronze	Collect like terms together										
Silver	Substitute into a formula										
Silver	Form equations										
Gold	Change the subject of a formula										
	Additional comments										
Year 7											
Year 8											
Year 9											
Year 10											
Year 11											

Functional skills

Difficulty	Objective	Year 7		Year 8		Year 9		Year 10		Year 11	
		Y/N	HT	Y/N	HT	Y/N	HT	Y/N	HT	Y/N	HT
Bronze	Work with time										
Bronze	Work with money										
Silver	Use the facts of proportion to scale up and down (e.g. recipes)										
Gold	Interpret data and figures (e.g. speed, distance and time, averages)										

Additional comments

Year 7	
Year 8	
Year 9	
Year 10	
Year 11	

Shapes and measures

Difficulty	Objective	Year 7		Year 8		Year 9		Year 10		Year 11	
		Y/N	HT	Y/N	HT	Y/N	HT	Y/N	HT	Y/N	HT
Bronze	Measure and estimate lengths and weights										
Bronze	Measure and calculate angles										
Silver	Read and apply scales on a map										
Silver	Calculate the perimeter, area or volume of shapes										
Silver	Draw the net of a shape										
Gold	Convert between units of measure										
Additional comments											
Year 7											
Year 8											
Year 9											
Year 10											
Year 11											

32

Number

Difficulty	Objective	Year 7		Year 8		Year 9		Year 10		Year 11	
		Y/N	HT	Y/N	HT	Y/N	HT	Y/N	HT	Y/N	HT
Bronze	Add, subtract, multiply and divide whole numbers										
Bronze	Express a quantity as a fraction of another										
Silver	Work with positive and negative numbers										
Silver	Express a number as a percentage										
Silver	Calculate percentages of amounts										
Silver	Share an amount into a ratio										
Silver	Add, subtract, multiply and divide with decimal numbers										
Gold	Calculate percentage increase and decrease										
Gold	Calculate the percentage change										
	Additional comments										
	Year 7										
	Year 8										
	Year 9										
	Year 10										
	Year 11										

Faculty support

The numeracy audit will generate lots of practical information that you can collate. Whether as an individual, a coordinator or a department, you will hopefully now understand the support given to numeracy in your subject area or across the full school. It is important that this information is assembled in such a way that it becomes useful data rather than just a paper exercise.

To do this, I centralise the comparable objectives for each subject area into year groups, as in the example on page 35.

This clear visual format is useful as a planning tool within the mathematics department. It enables staff to reorder the scheme of work or discuss ways of incorporating other subjects' application of the topic area, both to secure the foundations and to support other curriculum areas. As a result of completing the audit, some staff members realise how little they are doing with basic numeracy skills, but that these could easily be incorporated into certain lessons (especially after seeing the resources and ideas in Chapter 6).

The final part of the audit is a key component that allows staff to highlight their concerns on topic areas or weaknesses within their faculty:

> As part of the numeracy programme we are looking to provide support for faculties, particularly in light of some of the new specifications. As a faculty, are there any mathematical or numeracy topics you would require guidance on?

This creates an opportunity for direct discussions within the privacy of departments about their concerns and fears, but at the same time enabling you to support each other. It is also a very easy way for departments to highlight any concerns they have with the mathematical content of their courses.

The audit might highlight areas where the practice of specific skills is shared, such as drawing histograms in geography or comparing numbers written in standard form in science. It would be sensible to make sure the language used in the maths classroom is the same in geography and science classrooms. When students hear the same vocabulary and encounter the same methodologies being

		Add, subtract, multiply and divide whole numbers	Express a quantity as a fraction of another	Work with positive and negative numbers	Express a number as a percentage	Calculate percentages of amounts	Share an amount into a ratio	Add, subtract, multiply and divide with decimal numbers	Calculate percentage increase and decrease	Calculate the percentage change
PE	HT									
	Y/N									
Music	HT									
	Y/N									
Art	HT									
	Y/N									
Drama	HT									
	Y/N									
Business	HT									
	Y/N									
Computing	HT									
	Y/N									
Technology	HT									
	Y/N									
Religious education	HT									
	Y/N									
History	HT									
	Y/N									
Geography	HT									
	Y/N									
English	HT									
	Y/N									
Science	HT									
	Y/N									
Difficulty	Objective	Bronze	Bronze	Silver	Silver	Silver	Silver	Silver	Gold	Gold

used in various contexts, this will assist their long-term memory. It will also help teachers to deal with new topics to their courses.

The audit might also reveal areas that need clarity for both staff and students. An example of this might be the difference of interpretation of a line of best fit, which in maths means having an even amount of points on both sides of the line, and in science means going through as many points as possible.

Finally, the audit might draw attention to areas of mathematics that aren't explicitly covered within the GCSE mathematics curriculum, such as standard deviation or Spearman's rank coefficient. Students might be expected to use these in geography, but they may not have encountered them in mathematics. It may prove beneficial to do some research on these topics to enable you to provide each department with specific training and support. Another example would be to look into the types of graphs that students are expected to draw, use and interpret in science and design and technology.

There are usually different ways to solve a problem, so it can be useful to discuss with staff the various methods which can be used to arrive at an answer. This will include the students' calculator skills. Surprisingly, many students struggle to use a calculator effectively as they are uncertain about what each button does. They also fail to double check in case they have made a typing error or don't bother to perform a rough estimate of the answer so they know they are in the right ballpark. Some students and staff may need coaching to ensure they type in the operations on the calculator in the correct order. Knowing about and utilising standard procedures can also help to improve the consistency of students' mathematical understanding.

So far I know:

- How to start to tackle teachers' misconceptions about numeracy across the curriculum.

- The importance of carrying out a self-audit or whole-school audit of numeracy across the curriculum, of enhancing communication between departments and of sharing good practice.

- How to carry out an audit so an accurate picture of numeracy across the curriculum is gathered, but isn't a taxing demand on departments.

- That I have an example audit ready to be used or adapted.

- That I need to look outwards in the collation of data in order to gain the support of departments on their individual journeys.

The next stage is the introduction of the numeracy links. With these, teachers can explore in greater detail the categories they have seen within the audit.

Figure 4.1

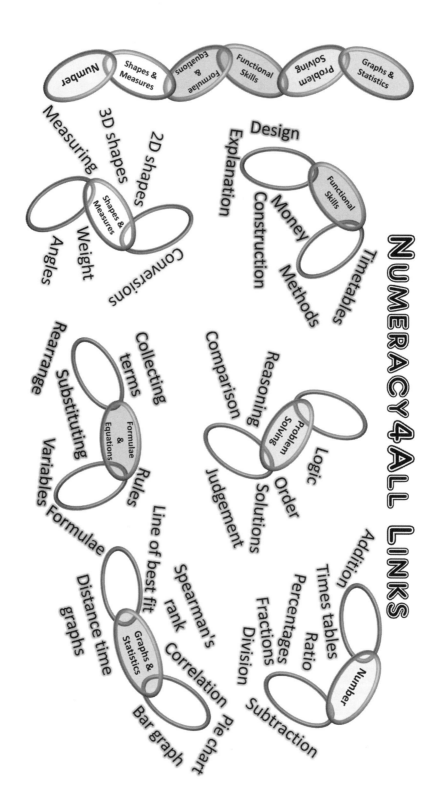

Chapter 4

Numeracy Links

The numeracy chain consists of six links that use a consistent method to demonstrate and embed numeracy across the curriculum. The links cover the six critically important numeracy skills which help to support students with their root level understanding of maths.

These links (see Figure 4.1) provide a comfort zone for teaching staff because they clarify what is expected when staff are asked to embed numeracy within a lesson. The links help to promote a balanced approach to planning and encourage teachers to enhance the lesson through the inclusion of some mathematical thinking. They also endorse a clear, consistent approach in methods related to numeracy. I can't emphasis enough the importance of having uniformity across a school when it comes to numeracy strategies.

Unpicking the links

In order to deliver a reliable and structured system, the school needs to adopt a collaborative approach to mathematical methods. For the process to work, all cogs need to turn in the same direction so the links mesh across the full school curriculum. Without an integrated approach, students may struggle if teaching staff are not applying the links or using the link skills as a fall-back method. This may then provide room for the students to make excuses for not using the approved methods. The more this happens, the more difficult it will be to establish this unified approach within the school culture.

The numeracy links have been designed from the student's perspective. The resources have been structured to promote personal independence in learning and to help embed consistent mathematical methods across the school. Given that a proportion of teachers may fear maths, the numeracy links aim to take some of the pressure off teachers having to remember mathematical techniques by providing them with useful prompts which can be used by both staff and students.

Each link comes with its own support mat. This enables students to have easy access to support documents in lessons. These could be displayed around the school or made into A5-sized booklets for use in the classroom.

Be aware that it is not always possible to define an activity with just one numeracy link; many activities will use multiple links. Frequently, the problem-solving and functional skills links will be used in conjunction with one of the other four links. The problem-solving aspect is the logical process underpinning the numeracy skills needed to find solutions to numerical problems.

For example, suppose you are in a food technology lesson tasked with designing a product to be cooked during the next lesson. This is a problem-solving task where you would need to chunk the challenge into several stages (Figure 4.2). This activity involves several of the link categories.

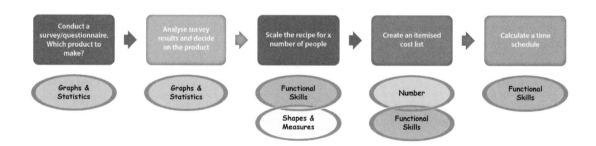

Figure 4.2

Problem solving

Problem-solving tasks are those with an open-ended nature – for example, buying the best mobile phone will depend on numerous factors such as the number of texts or the amount of data which is included in a phone contract. Puzzles, on the other hand, are closed in that they have one correct answer – for example, many maths problems as well as crossword and Sudoku puzzles.

Problem solving involves ordered and logical thinking to examine the information, formulate some kind of hypothesis and then test solutions to see how effective they are. Problems do not usually have a 'correct answer' but only 'the best answer so far' – think scientific research or the politics of educational policy. Sometimes it won't be obvious that finding a possible solution involves number work or completing a numeracy-related task. However, being able to fit together clues logically, predict likely outcomes and test a hypothesis are all forms of logical thinking linked closely to mathematical thinking.

Puzzle solving is the ability to bring order to chaos and come up with the correct solution. A good example of this would be a murder mystery activity. The students find or are given a number of clues which enable them to eliminate suspects based on logical reasoning. Treasure hunts work in a similar way. A key aspect of this link is being able to write a series of step-by-step instructions.

The problem-solving support mat (Figure 4.3) provides students with a series of generic prompts and questions to help generate new ideas and thoughts for when they reach a dead end. This helps to develop their higher order thinking skills and increases their independence.

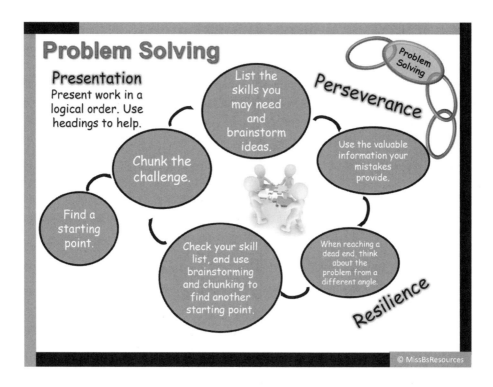

Figure 4.3 Download this resource from www.missbsresources.com/bookresources.

Graphs and statistics

Graphs and statistics are often the first things that come to mind for non-maths specialist teachers when they think of numeracy across the curriculum. Representing information graphically and analysing it statistically are key activities for topics in geography and science, but also extend to other subjects in which conclusions have to be drawn from data.

The graphs and statistics support mats (Figures 4.4 and 4.5) focus on a range of issues, from designing a questionnaire and selecting samples, to interpreting and analysing the data through the use of scattergraphs and bar charts.

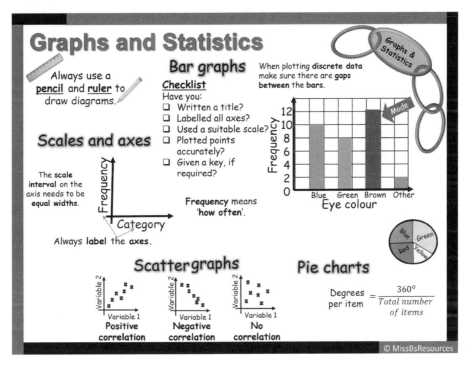

Figure 4.4 Download this resource from www.missbsresources.com/bookresources.

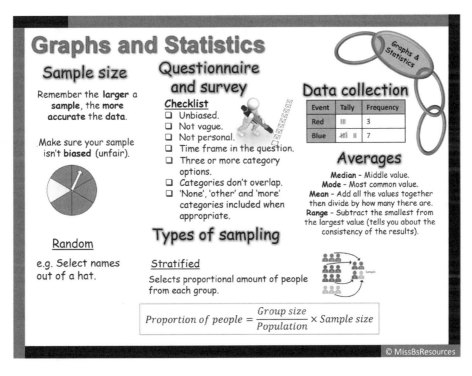

Figure 4.5 Download this resource from www.missbsresources.com/bookresources.

Formulae and equations

Formulae and equations are common in science, business and computing; however, they often appear in other subjects. One common complaint from maths teachers is that they have discovered other subject teachers teaching the tricks *behind* the maths instead of using the methods taught within the maths lesson. This can often confuse students and lead to them not remembering the knowledge because they don't understand where it comes from. A frequent topic where this transpires is with speed, distance and time.

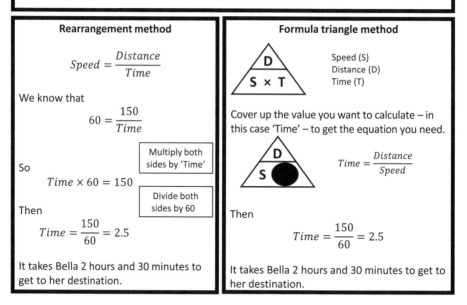

As shown in the example above, many teachers teach this module through the formula triangle method instead of the rearrangement method. This may give students only a superficial understanding of the topic as it acts as a substitute for making the link by changing the subject in common formulae. For the technique to become useful and memorable, it is important that students understand the reasoning behind the triangle method. Only when they understand the basic principle will students be able to generalise this approach and make links across the curriculum. Therefore, the rearrangement method should come before students are given the memory technique of the triangle. This has become more imperative in recent years, given the increased demand for students to be able to apply and rearrange several formulae and equations in a wide array of subjects.

The formulae and equations support mats (Figures 4.6 and 4.7) provide a range of examples and tips on how to solve, rearrange and substitute into equations. Their purpose is to remind students of the basics.

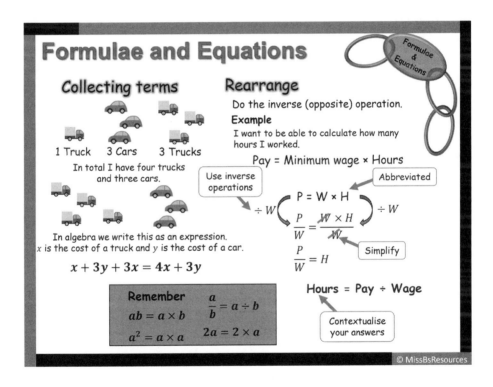

Figure 4.6 Download this resource from www.missbsresources.com/bookresources.

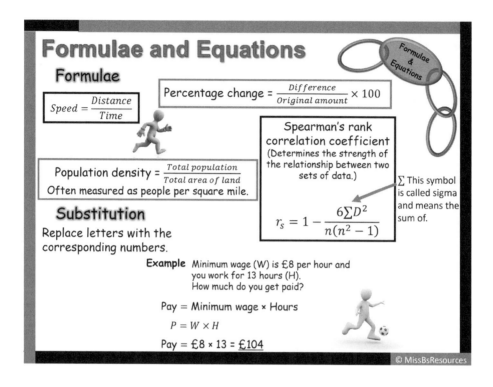

Formulae and Equations

Formulae

$$Speed = \frac{Distance}{Time}$$

$$Percentage\ change = \frac{Difference}{Original\ amount} \times 100$$

$$Population\ density = \frac{Total\ population}{Total\ area\ of\ land}$$

Often measured as people per square mile.

Spearman's rank correlation coefficient
(Determines the strength of the relationship between two sets of data.)

$$r_s = 1 - \frac{6\sum D^2}{n(n^2 - 1)}$$

$\sum$ This symbol is called sigma and means the sum of.

Substitution

Replace letters with the corresponding numbers.

Example Minimum wage (W) is £8 per hour and you work for 13 hours (H). How much do you get paid?

Pay = Minimum wage × Hours

$$P = W \times H$$

Pay = £8 × 13 = £104

© MissBsResources

Figure 4.7 Download this resource from www.missbsresources.com/bookresources.

Functional skills

Many students fail to grasp the practical relevance of maths and numeracy to their current and future lives. Functional skills are those practical abilities which allow people to gain the most from work, education and everyday life.

Nearly everything we do is governed to some extent by time, from making appointments and knowing how long it will take to do something or get somewhere, to remembering birthdays and school terms. These all require an understanding of both the 12-hour and 24-hour clock and to know the months of the year. However, there is a tendency for people to become deskilled as they come to rely on increasingly widespread mobile technology rather than their own thinking. But when the technology fails, they find themselves unable to tell the time and are then late for appointments or for work.

In general terms, functional skills involve recognising patterns and being able to formulate a theory or story about what is happening, being able to use information to make decisions about what to change and being able to evaluate the results of making changes. In practical terms, functional skills range from setting an alarm clock or calculating the quantities of ingredients needed to make a batch of scones, to being able to conduct an investigation to find the cheapest energy supplier or finding a way of improving a sports team's results (which might require being able to calculate and compare goal averages or the average speeds of runners). These functional skills are becoming an ever increasing part of the maths curriculum, with students having to apply their mathematical skill to contextualised real-life maths questions.

Although it is assumed that students will pick up these functional skills informally during their school careers and throughout their adult lives, this may not happen. Instead, some students learn how to fake these skills simply to get by or be left alone. If schools and teachers recognise that students have missed out on some of the basics and are now struggling, they need to take action urgently to remedy this situation. Teachers need to unite in terms of highlighting when functional maths skills are required, using consistent methods and indicating how these should be used by the students in a particular context. If these numeracy and mathematical skills are clearly identified – that is, if teachers help them to connect the dots – this greatly supports and encourages the students' willingness to learn and develop their numeracy and mathematical capability. As a result, the students will rank the importance of the subject and the skills more highly *and* realise that they 'can do maths'.

One problem that many maths teachers struggle with is the 'Why do I need this?' question. If you simply tell students about the importance of being numerate in everyday life, many of them won't believe it. It is far better to get the students to explore the issue for themselves. You could ask them:

- Have there ever been occasions when you wanted to do something but couldn't because you were unable to work out what to do or how to do it?

- Are there situations you avoid because you think you'll be taken advantage of?

- Have you ever cheated or paid the wrong amount for something?

Figure 4.8 Download this resource from www.missbsresources.com/bookresources.

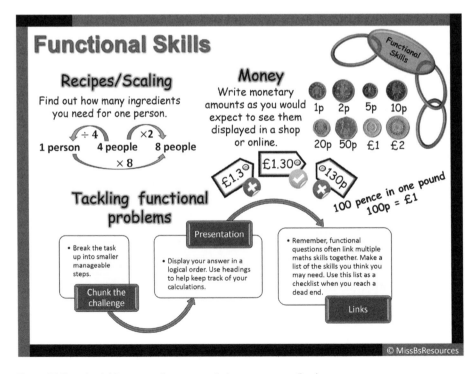

Figure 4.9 Download this resource from www.missbsresources.com/bookresources.

In other words, get them to draw on their own experiences of getting things wrong, being confused or paying over the odds, and then suggest that there is a way they could avoid this in the future. This approach should be amplified with assistance from other subject teachers across the whole school. Opportunities should be found for students to try out various numeracy strategies in safe environments. Actions really do speak louder than words, and seeing numeracy in action in real-life scenarios is vital.

Functional skills are those that the students will be using on a daily basis to survive. This includes both simple and complex tasks – from being able to read a timetable and calculating how much change they need, to being able to interpret data, averages and ratios.

The functional skills support mats (Figures 4.8 and 4.9) focus on the basic skills of working with time, money and scaling.

Shapes and measures

The ability to understand various measurement systems or to calculate quantities is a vital skill that all students require for shopping, planning a holiday or working as a builder or hairdresser.

Using shapes and measures is most common in design and technology subjects – for example, when looking at packaging and nets, designing products and drawing images. But students also need to understand how measurements are calculated – for example, when we buy a carpet we need to be able to calculate how many square metres are needed to cover a given floor area, or when buying a piece of furniture so that it will fit in the designated space.

The shapes and measures support mats (Figures 4.10 and 4.11) focus on methods for finding the area, volume and nets of shapes, as well as looking at the use of measuring instruments and key conversions.

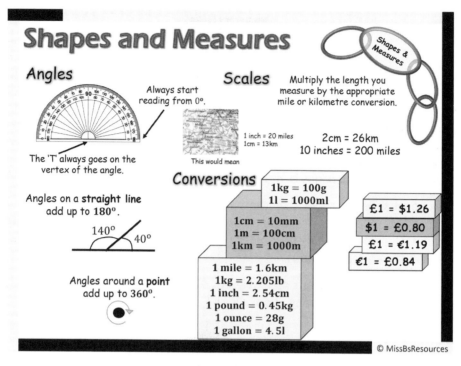

Shapes and Measures

Shapes & Measures

Angles

Always start reading from 0°.

The 'T' always goes on the vertex of the angle.

Angles on a **straight line** add up to 180°.

140° 40°

Angles around a **point** add up to 360°.

Scales

Multiply the length you measure by the appropriate mile or kilometre conversion.

1 inch = 20 miles
1cm = 13km

This would mean

2cm = 26km
10 inches = 200 miles

Conversions

1kg = 100g
1l = 1000ml

1cm = 10mm
1m = 100cm
1km = 1000m

1 mile = 1.6km
1kg = 2.205lb
1 inch = 2.54cm
1 pound = 0.45kg
1 ounce = 28g
1 gallon = 4.5l

£1 = $1.26
$1 = £0.80
£1 = €1.19
€1 = £0.84

© MissBsResources

Figure 4.10 Download this resource from www.missbsresources.com/bookresources.

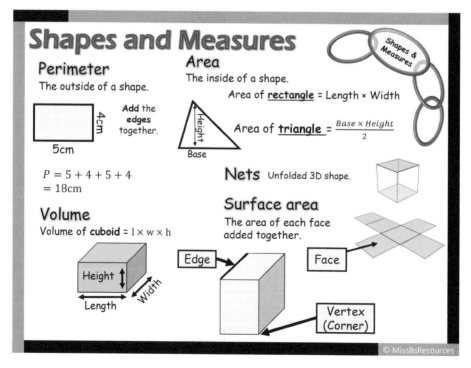

Shapes and Measures

Shapes & Measures

Perimeter

The outside of a shape.

Add the edges together.

4cm
5cm

$P = 5 + 4 + 5 + 4$
$= 18$cm

Area

The inside of a shape.

Area of **rectangle** = Length × Width

Height
Base

Area of **triangle** = $\frac{Base \times Height}{2}$

Nets Unfolded 3D shape.

Volume

Volume of **cuboid** = l × w × h

Height
Length
Width

Surface area

The area of each face added together.

Edge

Face

Vertex (Corner)

© MissBsResources

Figure 4.11 Download this resource from www.missbsresources.com/bookresources.

Number

The fundamental process of numeracy is the ability to calculate numbers using basic addition, subtraction, multiplication, division and order of operation rules. It also extends to the more complex skills of finding fractions and percentages of amounts.

Almost all situations in working life demand basic number skills; students need them if they are to hold down even the most basic of jobs. Without a good understanding of numbers, they can be putting themselves or others in danger (e.g. in measuring out dosages of medicine), or simply losing out or mismanaging finances through being incapable of calculating the best value for money.

When working with teachers, I have found that there is often confusion over the methods students should be using to complete the basic four operations. Some methods go in and out of favour, and teachers and parents may find that things are very different from their day. As a result, they can feel intimidated by the most basic of maths problems because they don't understand the methods the students have learned.

The number support mats (Figures 4.12 and 4.13) provide staff with some common methods used for the basic skills, as well as those of a more complex nature. The intention is to establish a standard approach so that both students and teachers have a common method for performing these basic operations. The support sheets focus on topics which are commonly seen in subjects across the school, such as being able to calculate a percentage of a score.

Embedding the links

The numeracy links are designed to be used in every teacher's classroom practice. The use of numeracy in a lesson should always be content specific to the lesson, enhance the lesson/task or help the students to engage with a task. Numeracy shouldn't be tagged on as a last-minute afterthought. The numeracy link support mats are designed to work as posters for display or to be made into booklets to go on classroom desks. When numeracy tasks are incorporated within a lesson,

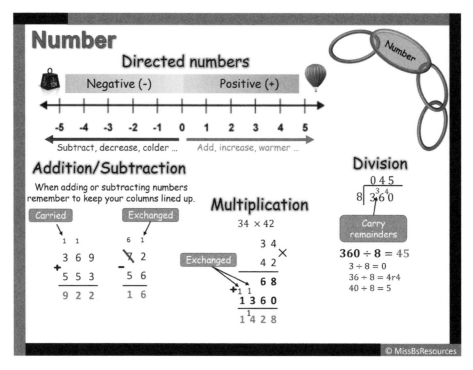

Figure 4.12 Download this resource from www.missbsresources.com/bookresources.

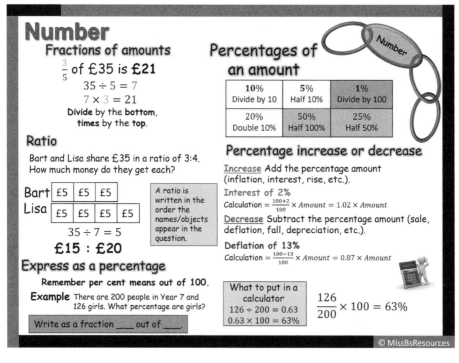

Figure 4.13 Download this resource from www.missbsresources.com/bookresources.

best practice is to include the relevant numeracy links on hand-outs to direct the students to which support documents to use independently, if needed.

The launch of the links in a school or your own classroom should be planned carefully. A good way of doing this is to focus on showcasing and celebrating the impressive work already happening in your classroom or any other classrooms in the school. In order for the links to be successful, knowledge about them needs to be embedded with both staff and students. Any wavering or inconsistencies in the use of the links could very easily cause the system to fall apart. Remember, a chain is only as strong as the weakest link – and that includes the Numeracy4All chain.

For the initial launch of the project, I've found it beneficial to invite staff to carry out a detailed inspection of the new link topic areas and support mats in their own departments and faculty groups, so they can see what the numeracy links are all about and how they might be useful in supporting their particular curriculum area and current schemes of work. A good way of providing hands-on practice for staff in how to use the links is to model a scenario where the mats could be used and ask staff, in small groups, to complete the task through the students' eyes. It is advisable to provide Q&A sessions to clarify any issues that arise.

Any training and development with staff should start with this kind of practical approach. Teachers need time to familiarise themselves with the numeracy link support mats if they are to support the students. They also need time to evaluate and discuss how they could embed this within their own practice. The more they share ideas for best practice with each other, the better they will be in enhancing learning across the whole school. This discussion time will be invaluable to boosting the confidence of staff in using the resources in the classroom.

Sharing the links

In one of my previous schools, I introduced the numeracy links during faculty meetings. I reserved the whole meeting to present resources and ideas and to give teachers an opportunity to share their good practice. This allowed me to tailor the delivery of the links to each faculty; as you can imagine, French was very different from design and technology.

By placing the links logo on relevant class materials, the students are encouraged to become more independent in solving their own problems. A benefit for the teacher is that it frees up time for them to work with those who need extra support or stretching. It also enables the students to get into the right mindset to link their numeracy skills in different lessons across the school.

As with any new method of teaching or school initiative, the students need to be made aware of the intended purpose and outcomes, and how best to achieve them. Giving students model tasks in form time or lessons where they have a chance to explore the links – similar to the teacher training – will allow them to become familiar with how to access the support available to them. It is key that they are aware of how numerate they need to be on a daily basis, as this helps to drive up standards and stress the importance of numeracy to them personally. Discussions during form time can be an easy and non-pressurised way to encourage the students to discuss numeracy openly. The more students realise the impact that numeracy skills have on their lives, the greater value they will give it.

So far I know:

- The numeracy links are a physical aid to highlight numeracy across the curriculum.
- Unpicking the links exposes the six key areas:
  1. Problem solving
  2. Graphs and statistics
  3. Formulae and equations
  4. Functional skills
  5. Shapes and measures
  6. Number

- When to potentially use the links as visual prompts in the classroom.
- How to embed the links across the school as part of the school's numeracy across the curriculum development plan.
- When to share the links with staff and students.

Calculate 25% of £30

Method 1

100%: £30
50%: £30 ÷ 2 = £15
25%: £15 ÷ 2 = £7.50

25% is worth £7.50

Method 3

100%: £30
10%: £30 ÷ 10 = £3
5%: £3 ÷ 2 = £1.50
25%: £3 + £3 + £1.50
= £7.50

25% is worth £7.50

Method 2

100%: £30
25%: £30 ÷ 4 = £7.50

25% is worth £7.50

Method 4

0.25 × £30 = £7.50

25% is worth £7.50

Method 5

$\frac{25}{100}$ of £30   (30 ÷ 100) × 25 = £7.50

25% is worth £7.50

Chapter 5
Subject Knowledge

Making connections

The skills that students need for being numerate are taught primarily in the mathematics classroom. As we have seen, these skills are best remembered when they are continuously secured and enhanced across the full school curriculum. The numeracy links are the starting point in getting the students to make connections and use their maths skills in multiple subject areas. The dream vision is for the students to make connections between the subject areas unprompted.

Staff should be able to recognise when a student is struggling. They might do this by ascertaining the level of the student's subject knowledge and the methods they are using to solve problems, as these may be different from the preferred method. Alternatively, if numerous students are struggling with a particular topic or method, then staff across the school may need to standardise the way things are done. In this way, everyone becomes familiar with the recommended whole-school method.

Problems can arise because there are usually a number of different methods for performing certain mathematical operations, such as calculating percentages. I am of the opinion that students should be allowed to solve problems using the method they are most comfortable with – as long as it is effective and transferable. If they are making mistakes or the method isn't transferable to harder problems, then students need to be willing to use and learn the school's standardised fall-back method. (For example, see page 56.)

These are just a sample of the methods that can be used to calculate a percentage of an amount; there are many others. All of the above are perfectly valid methods; however, method 4 is the optimal method when a calculator is available. Students can use alternative methods to the method of choice if the calculations are easy; however, when they are finding things difficult, it is important that teaching staff promote a consistent standard method.

It may be useful for students to discuss the variety of methods they might use to tackle the numerical part of the task. This will help weaker students to make the connection between the mathematics classroom and the task being asked of them. Bearing in mind that many Key Stage 3 classes are of mixed ability, it is helpful for students to have 'numeracy buddies', as it's often far more effective for them to look to their peers for support. The numeracy links can stand as a second line of approach, and then the classroom teacher if all else fails. It is important for students to gain numerical independence and appreciate the importance of mathematical skills.

Students need to be able to calculate sums and estimate answers in their head as well as understanding how to use a calculator. It's important to provide students with scenarios where calculators aren't available for use. When a student struggles with mental arithmetic, it is not good enough for them to counter with, 'I'll just use my computer/phone', because they will not be sufficiently aware of the kind of answer they should get, especially if they mistype any inputs. Remember that technology is only as good as the information we put into it.

A practical application

A good example of putting learning into practice is when a student later becomes a home-owner and needs to redecorate. They will need to be able to calculate surface areas in order to know how much paint or wallpaper to buy. To solve this kind of everyday problem, they need to be able to think logically, add, multiply, divide and understand basic shapes.

For example: a litre tin of paint costs £6.99 and covers an area of $8m^2$ of wall space. How much will it cost to decorate a room which has the following dimensions: 8m × 4m × 2.5m (for the sake of simplicity we are ignoring the door and windows)?

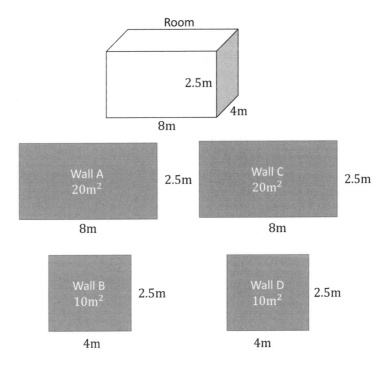

The room has four walls and each wall has its own surface area. The total area of the walls is 20 + 20 + 10 + 10 = 60m².

The students need to work out how many tins of paint they need, so they calculate the area of the walls divided by the area the paint covers: 60 ÷ 8 = 7.5 tins are needed. Unfortunately, they can't purchase half a tin so they will need 8 litres in total. This will cost 8 × £6.99 = £55.92.

This may seem straightforward, but some students struggle with these kinds of problems and are unable to generalise with different quantities, costs and so on. Although the problems may be similar in structure, they are thrown by the variations and ensuing calculations, so it is important that the students understand the basic principles of problem solving.

Teachers can support their students by establishing a culture in which numeracy and basic numerical ability are an essential part of the conversation. None of the students can opt out of practising these skills; taking part is non-negotiable.

In this way, every student comes to understand the value that every teacher is placing on numeracy. Consequently, some students may feel more comfortable as they are less under pressure when practising the basics outside of the mathematics environment.

Mentors and numeracy leaders

Numeracy leaders are students who are confident in mathematics and who can support staff and other students across the school in any subject area. They can be called on if there is a moment of uncertainty in a lesson (e.g. which method to use) or to support other students who have some kind of mathematical problem. Numeracy leaders can also be used to mentor underachieving students in lower year groups; these students become mentees. I feel it is important for the mentees to be from year groups lower than the numeracy leader to help avoid intimidation. The mentoring programme also has the bonus outcome of providing some of the more vulnerable Year 7 students with an older student as a role model – a friendly face from Year 9 or 10.

To become a numeracy leader, a student should be working above the national average according to their Key Stage 2 SATs scores. Getting students to act as numeracy leaders is a great way to challenge any student who is more able at mathematics by supporting weaker students who haven't grasped it yet. It is every teacher's dream to have the time to provide one-to-one support for all students who may be struggling with core concepts. By creating numeracy leaders you can offer this one-to-one support. There is also the potential for numeracy leaders to gain a leadership qualification. It is worth looking into official leadership qualifications as these will stand those students in good stead in later life. It also adds an extra level of commitment.

When selecting student numeracy leaders, consider not only their mathematical ability but also their punctuality, reliability and interpersonal skills. They need to have good communication skills, the ability to coach their mentee rather than simply telling them what to do and they should not make their mentees feel intimidated by the maths. Each numeracy leader will have an individual mentee assigned to them, so they need to show a genuine commitment to supporting them when the going gets tough. When setting up the programme, you may

Chapter 6

The 31 Prime Resources and Ideas

It is important that numeracy is integrated into a lesson rather than just being an add-on or afterthought. The best numeracy-related tasks are those that are naturally imbedded within a subject area – for example, graphical representations in geography, nets in design and technology or standard form and formulae in science. This doesn't mean that you can't include numeracy-based activities in a traditionally non-numerical subject; however, numeracy should have a relevance and purpose according to the topic being taught.

Any numeracy across the curriculum resources you decide to use should be chosen according to the criteria of whether it will enhance the learning. If it will detract from the objective of the lesson then consider carefully whether it is worth the distraction. Please don't just include numeracy resources to tick a box on an observation spreadsheet. There will be many opportunities over time to explore numeracy content and these lessons should be planned in advance.

The following generic numeracy resources and ideas are designed to work in any classroom. They cover a range of activities and are intended to boost the learning and engage the learner. They also have the added bonus of including numeracy content to help support students' numerical foundation.

The resources are divided into the following categories:

- Literacy

- Exploration

- Engagement

- Marking and reflection

- Organisation and presentation

- Classroom management

Many of the resources need some preparation beforehand. In practical terms, you will need to create presentation slides or hand-outs. It is useful to print and laminate generic resources (e.g. Resource 2 – Writing Weigh-In) so that you have them in your room and ready to be used.

Throughout this chapter, I have provided some basic mathematical pointers, such as the language to be used and the numerical concepts that students often struggle with or have misconceptions about. You may find it useful to have a maths dictionary available in your classroom to encourage students to refresh their memory independently.

When students are completing any numeracy-related task and need help, it may be beneficial to use some of the following questions to help prompt their understanding.

- **Can you see numbers similar in size to these that are easy to deal with?**

  This question allows students to estimate their answer. For example:

  $$2 \times 6.9$$
  $$\approx 2 \times 7$$
  $$\approx 14$$

  $$2 \times 6.9 = 13.8$$

  The question may also prompt them to calculate the answer in a different way. For example, students will often struggle to compute 99×7 in their heads. However, they can easily work out $100 \times 7 = 700$ and then subtract one lot of 7: $700 - 7 = 693$.

- **Could you break the question up and work on each part separately?**

  It is important that students chunk a challenge. For example, if they are being asked to cost an event in order to work out the profit, they may be given a lot of information all at once, such as the price of adult and child tickets, how many adults and children are predicted to attend and the cost of staging the event. They could break this question up by following these steps:

  > Estimate the total of adult ticket sales.

  > Estimate the total of child ticket sales.

  > Estimate the total of the ticket sales.

> Work out if they have made a profit by deducting the cost of the event from the ticket sales.

- **Would using a number line help?**

  Number lines can help the weakest students when they are adding numbers using the bounce method. Number lines are particularly useful when the students are asked to deal with negative numbers or to find the difference between two numbers.

- **Can you use your knowledge of quarters, halves and doubles to help?**

  To find 50% most students would be able to halve a number, and to find 25% they often learn to halve and halve again. Most students are comfortable with calculating using this technique.

For each resource I have included the symbols from the numeracy links to indicate the territory covered by the resource.

The bounce method

Example: to find the difference between 163 and 198

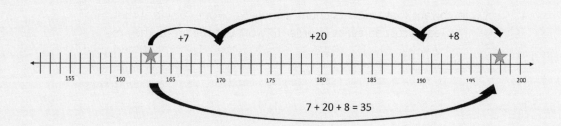

- To calculate the difference using the bounce method, the students first need to locate their numbers on the number line.

- They then establish how many ones/units it takes to get to the nearest 10 (in the example, this is +7).

- Then they add on the tens (2 × 10 = 20). This should get them as close to their required number as possible, but not over it.

- They should then add on the extra ones/units (+8).

- Finally, the students add together all the numbers they have collected to find the difference (= 35).

Software

Some resources will need specialist software which can be downloaded (for free) from the Internet. Sometimes you will have to register to use this. There will also be occasions when the students will need to access their own mobile technology, but most of the time this will not be necessary.

Copies of the resource printouts can be downloaded as pdfs from: www.missbsresources.com/bookresources. It is free to register on the site to access these.

Literacy

1. Scrabblecross

Scrabblecross is a relatively closed task but with many avenues for the students to go down. Problem solving and subject knowledge are required for this task.

Preparation and resources

Choose two to five key words and make a crossword puzzle out of them. There is free software available for doing this as well as many apps. However, I have found the most useful one to be www.puzzle-maker.com/CW.

How it works

1. Create a crossword out of three to five key words.

2. The only clue to be given to students is the topic, source, text or case study.

3. Provide students with the letter tiles that make up the words and the total of each word.

4. Students then need to apply their knowledge of the subject to solve the puzzle. A tile cannot be used twice.

A nice twist on this activity is to allow students to make up their own puzzles and swap them. This encourages them to include key words that they think their partner won't know. This exercise will extend their vocabulary and knowledge.

To differentiate this activity you could provide clues or questions to one or all of the words.

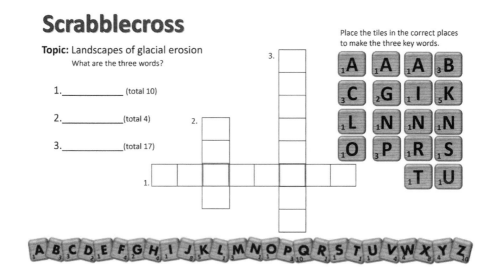

Scrabblecross

Topic: Landscapes of glacial erosion

What are the three words?

1._____ (total 10)

2._____ (total 4)

3._____ (total 17)

Place the tiles in the correct places to make the three key words.

Answers: (1) Abrasion, (2) Tarn, (3) Plucking.

Activity types

Starter/plenary

Example activities

PE – When students are looking at key terms for different movements and positions within sport. A great way to focus the students' minds and refresh their knowledge.

Computing – When asking students to study Internet safety, provide students with a Scrabblecross and ask them to research the issue on the Internet. As the students find the key terms they will need to check if they fit into the Scrabblecross given to them. This will help to keep them on track and focused. The key words could be *password*, *report* and *SMART* (a common acronym used when teaching e-safety).

English – When learning grammar, warm students up by arranging the grammar terms into a puzzle.

Mathematical tips

Sum and *total* mean to calculate the value by adding the numbers together.

2. Writing Weigh-In

Some students have a poor understanding of weight, particularly when it comes to converting from one system of measurement to another. This activity helps students to gain a better understanding of weight through applying VCOP (vocabulary, connectives, openers, punctuation) to improve their extended writing skills. The activity also gives students a chance to practise the basic skill of addition.

To do this, weight is assigned to various aspects of an activity. This could be for writing using correct grammar or for constructing an essay. The more sophisticated the answer, the heavier the assignment. The weigh-in specifically focuses on the quality of the content rather than the length of the assignment.

Preparation and resources

Each student should be given a Writing Weigh-In mat (this can be laminated for reuse) and an appropriate initial target weight for their essay. Students should be comfortable in peer and self-assessment to help reduce the initial teacher workload.

How it works

1. Set the minimum weight for the assignment, such as 1kg in weight. This can be adjusted for different groups of students.

2. Using the Writing Weigh-In mat, students should initially peer or self-assess their work in order to calculate the initial weight of their assignment.

3. Next, the idea is to find ways of making improvements to their writing to increase its weight.

4. Set the minimum improved target weight for the assignment. State this in terms of kilograms rather than grams so the students will need to use their scaling/conversion skills. As before, the target weight can be adjusted for different groups or individual students.

5. Students should note any categories which are under-represented in their work.

6. If improvements need to be made to reach the new target, then support should be given in how to write complex sentences. If students achieve their target, then a new target weight should be set.

The quality of the assignment should improve as students use more complex sentences instead of simple sentences and key vocabulary instead of vague language.

The higher the level of content a student uses within a category, the more weight they add to their essay. For example, if students use basic coordinating conjunctions such as *and*, *but* or *nor*, they would receive 50g of weight. However, using subordinating conjunctions, such as *after*, *although* or *whenever*, would add 300g of weight to their essay.

Activity types

Marking and feedback

Support sheet

Extended writing

Example activities

English – When providing feedback on an extended piece of writing, you could ask for students to improve the weight in grams of their essay by *x* amount or *y* per cent.

History – In preparation for examination questions, support mats could be given to students as a support tool to help them become more independent and consistent in their approach to the questions and structure of the assignment.

Mathematical tips

Remember that 1kg = 1000g and 0.5kg = 500g.

Teacher notes

Some students are only able to use capital letters and full stops, whereas others are more ambitious in their writing and use a wide range of punctuation, including commas, semicolons and colons. Students need to be able to demonstrate the correct use of higher level and more ambitious punctuation to receive the highest possible marks in their GCSEs.

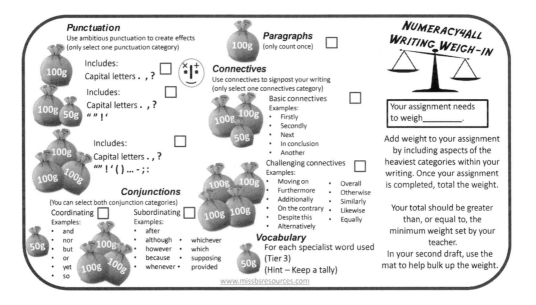

Variations

The concept of the weight of an answer could be adapted to be subject specific. For example, if a student is writing a newspaper article then you will need to identify the key defining features that would add weight to the piece of work. When designing a support mat like the one above, it is often easier to work backwards by starting with what a good example would look like. You should also consider three different standards of work (on, above, below) when assigning weightings to the different key characteristics. Bonus weight categories for writing a newspaper article could include:

- Addresses the standard criteria of who, what, when, where, why and how.

- Starts with a leading sentence.

- Written in the past tense.

- Includes subheadings.

- All information relates to the main idea.

- Gives all the important details.

- Follows up main facts with additional information.

- Concludes the article.

3. Mathematical Language in Extended Writing

Many mathematical terms are used in everyday speech – for example, 'back at square one', 'we have come full circle' and 'part of the equation'. Using mathematical language in sentences outside of a typical maths environment can encourage students to understand the meaning of each word so they can apply it in the correct context and help them to broaden their range of vocabulary.

Preparation and resources

Beforehand, you will need to devise a potential list of mathematical or scientific words you think would be appropriate to the activity for students to use as a prompt.

How it works

1. Provide students with a grid or list of mathematical words that are relevant to your subject area. You may find it interesting to add a few mathematical words to the list for which there isn't an obvious link and see what the students come up with. I love to be surprised and sometimes find I have missed the obvious!

2. The students are given the list and then attempt to use as many of the words as they can in relevant contexts in their writing to show how well they understand the meanings of the words. This also helps to develop their writing skills as there are many ways in which maths vocabulary could be brought into the equation in an open-ended writing task.

Some mathematical words you may wish to use might include:

addition	greater than	binary	denominator
subtraction	less than	congruent	equilibrium
product	average	constant	extrapolate
proportion	equal	correlation	frequency
quotient	equivalent	degree	improper
ratio	fractal	derivative	infinite
sum	statistical	factor	iteration

Activity types

Extended writing

Example activities

Drama – Ask students to direct each other into position using mathematical language (i.e. geometry coordinates) or to use words such as 'ratio' or 'proportion' to get them to move into different sized teams or groups.

English – When students are asked to make comparisons between characters and themes they could, for example, comment on the frequency of a character's appearance or the repetition of a particular verb or phrase. They could also explore factors which might have caused an event to occur.

Mathematical tips

Have an online mathematical dictionary available for those students who want to look up the definitions of the words.

4. Talk Time

How good are the students at talking in front of a group without any long pauses or fillers such as 'umm', 'eerr' and 'aahh'? Communication skills are a key aspect of many occupations, and as teachers it is part of our job to make sure our students acquire and practise the skills of public speaking for those times when they want to get their ideas or point of view across in an effective manner and make a good impression on other people.

Preparation and resources

You will need to have timers available for students to use. If they are allowed to bring their own devices into school, I often find it easiest to trust the class to use their own phones, especially if I want to run this as an on-the-spot activity.

Decide whether to give students preparation time or not for this exercise. I opt to write key topics or themes on the board and then give students the first five minutes of the lesson to make notes on some of the topics and themes. This gives me time to complete the register, as well as giving the students a chance to revise the notes they have previously made in their books.

How it works

1. Communicating fluently is vital in this activity. In pairs, the students give short talks on selected topics from a list on the board and time each other when doing so.

2. If a student pauses for more than two or three seconds, resorts to 'umm', 'eerr' or 'aahh' when speaking, goes off topic or repeats themselves, then the listening student stops the timer. They then change roles.

3. If the second student is stopped they again change roles, but this time they move on to the second topic and continue like this until all the topics have been covered.

4. Each time they change, the listening student should note down how long the speaker was speaking for.

5. This cycle continues until either all the topics have been used or you stop the activity. At this point, give the students time to add up their talk times and calculate how long in total they were able to talk for.

Activity types

Speaking and listening

Starter/plenary

Example activities

Science – Students talk about different types of chemical or biological reactions.

Drama – Choose standard drama topics for students to improvise on.

Music – Ask students to discuss different genres of music.

Mathematical tips

Refer to the functional skills support mat on working with time (Figure 4.8).

Variations

Students could plot their timings over the term to see how they are improving. You could start a group discussion about which strategies work best. Alternatively, you could treat this as a competition so that the person who was able to talk the longest is the winner. You could also use the data from the whole class as the basis of further maths or statistics exercises.

5. True Value

True Value asks the teacher to assign a monetary value to words: less important words have a low value, while important key words in a topic have a higher value. When the students are writing a piece on a topic, using high value words would suggest a greater understanding.

So, if you are going to price your writing, how can you maximise the value? Who will be the richest student?

Preparation and resources

You will need to prepare a list of key words that you would expect to be included in an essay on a particular topic. Each word is assigned a monetary value which bears a relation to how important it is within the context of the essay.

When students have to write about a topic or explain a concept it can often be hard for them to know where to start. By providing the students with a grid of key words (or characters, dates or events) to which you have assigned a value, it should be easier for them to structure their writing to create maximum value. The students will still need to arrange the key words in a sensible way; the key words merely act as signposts for achieving the top value to their writing.

Here is an example from a music lesson:

Timbre	Crescendo	Staccato
£4	£1	£2
Syncopated	Texture	Forte
£4	£3	£1
Canon	Countermelody	Melodic
£4	£3	£3
Diminuendo	Adagio	Allegro
£1	£3	£2

How it works

1. Project the table of key words onto the board and ask the students to identify which words would be most useful for logically structuring their writing.

2. Then ask them how they could connect the remaining words into their plan.

3. Students need to use the key words accurately and with purpose. It is at the teacher's discretion whether to penalise a student for inserting all the words with little thought of context just to gain value.

4. To make the activity more challenging you can place a minimum value that the writing needs to achieve.

Activity types

Extended writing

Planning

Plenary

Example activities

Music – Ask students to describe a piece of music.

History – Use as a writing prompt with key dates, figures and events.

MFL – Use a list of verbs that will tell a story.

Mathematical tips

See the functional skills support mat (Figure 4.9) for advice on calculating with money.

Variations

For higher ability students, use the currency from the languages they learn in your school (e.g. euro, yuan). As an extra extension, provide students with a mixture of amounts in both sterling and the country's currency, as well as an up-to-date exchange rate. (There are many websites which provide monetary exchange rates, e.g. www.xe.com/currencyconverter/.)

Exploration

6. Weight of the World

The weight of the world is resting on the students' shoulders. This is often the case when students first read a question: they will have many different and disconnected ideas running through their heads. Therefore, it is important that they sift through this information and choose the ideas which best link to the topic. To stop the world from toppling over with their abundance of knowledge, students need a secure structure. To do this they have to connect as many of the information boxes together as possible by creating bonds between them.

The focus of Weight of the World is on the collection of data using both tallying and frequency tables. However, the key to this activity is to make multiple connections and links between key terms, facts and knowledge. This is especially useful as a revision strategy.

Preparation and resources

You will need to provide five to ten key words or phrases (e.g. people, places, ideas, dates) relating to a specific topic, article or case study for students to make connections between. This activity can be carried out in exercise books but I often find it is easier if the connection chart is drawn on a blank sheet of A3 paper.

How it works

1. It is up to you whether you supply all the key words, a selection to choose from or allow students to compile their own list of words.

2. Students need to create as many bonds as possible between the key words to stop the world from toppling over. The aim is to make at least 25 bonds. Students get a bonus of one bond every time they make three connections from the same key word.

3. When connecting two key words, students should write the reason for the bond on the line.

4. They should then peer mark in groups, discussing the bonds they have proposed and whether they should be accepted.

5. Students should tally and total their bonds and bonuses in a table.

6. The student with the highest score wins.

7. Afterwards, students can confer with one another. If a student doesn't have a particular bond, then another student should suggest one which they should add using a different colour.

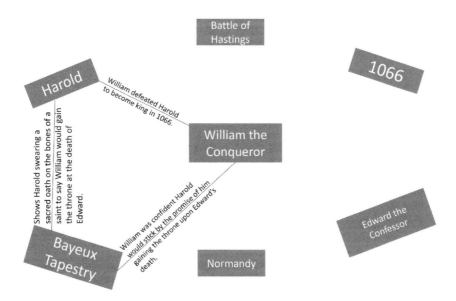

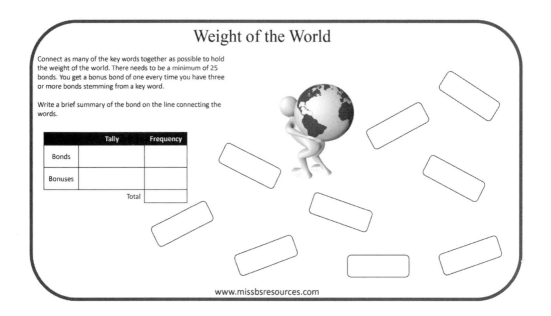

Activity types

Exam questions

Coursework

Extended writing planning

Example activities

Geography – When using case studies, select a number of key terms relevant to the topic being studied and the case study. Ask students to make connections based on an examination question.

Music – When planning a composition, students should be looking to meet key characteristics of the brief. These characteristics could include, for example, major, minor or other chords, particular tempos, instruments and genres. Students should then link the terms together, either through musical motifs, instrumentation or key changes to compose a piece of music to represent, for example, 'A stormy night'.

Mathematical tips

When tallying information each line stands for one – for example, ||| represents 3. When a student gets to five they should put a diagonal line through the four vertical lines:

Frequency means how often – how many times did the event occur? This is found by totalling the tallies.

Problem Solving

Number

7. Impact Line

The Impact Line is based around a debating topic for which the students state arguments for and against different points of view. This can become a stimulus for group and whole-class discussions, thereby developing the students' reasoning and debating skills.

This exercise provides students with the opportunity to work with positive and negative numbers on a number line and with numbers in a context. This will help to deepen their mathematical comprehension of number.

The number line is the foundation of modern mathematics. Young children become fluent in using positive numbers as they learn to count. What requires more cognitive understanding is what happens when the number line is extended beyond zero into the realm of negative numbers. Understanding that negative 5 is less than negative 1 is a crucial mathematical concept that many students struggle to grasp.

Preparation and resources

You will need to provide each student with a number line sheet (see page 89). This can be laminated for reuse. Students could also draw a number line on a large piece of paper or a mini whiteboard.

You may also want a topic sheet which outlines a situation relevant to your subject. However, when the Impact Line is employed as a revision tool, students may use their books instead to find the relevant information.

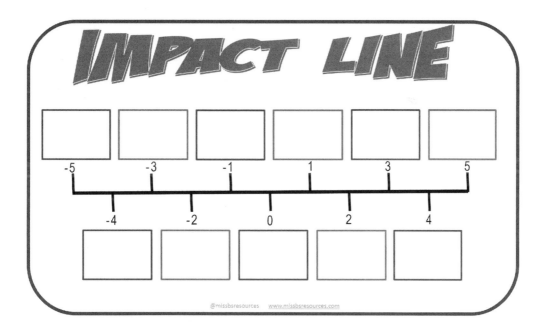

How it works

1. Students work individually to create a list of five advantages and five disadvantages, positive and negative effects or for and against arguments on a particular concept or topic. To speed up the activity, instead of asking students to create the statements, the teacher can provide them with statements or arguments to use as starting points for their discussion. Make sure the students spend sufficient time thinking about the issue and ask them to write down the key points before they discuss it with others. It is important that they are able to verbalise the reasoning behind their position and why any changes were made.

2. The students then work in groups of two or three to discuss where they think the statements should go on the Impact Line. They should justify their decisions about which statements will have the biggest or smallest impact.

3. The 'for' statements should be placed at the positive end of the number line and the 'against' statements at the negative end (e.g. positive 5

represents the largest positive advantage and/or best argument, while negative 5 represents the largest negative disadvantage and/or worst argument). It is important to make a point of getting students to discuss what the numbers represent as well as the order of the comments along the line. Students will often question what the zero represents. This could be a statement that has no impact or relevance to the topic, or one which could have equally weighted positive and negative impacts which cancel each other out.

4. Students should make notes throughout the group and class discussion on the points discussed. This could be done by annotating the number line and noting the increased or decreased value of a statement after discussion. For example, a student may have originally placed a statement at positive 3, but after discussion believes it now needs to move to positive 1. This is a change of -2. All of this information can then be used to help students answer an essay question or the higher order questions on exam papers which ask candidates to compare and contrast.

Impose a strict time limit on this exercise, as this should be the stem which leads on to other activities. However, it is important to make time for students to summarise the final position of the marker on the number line and describe how much movement there was during the debate.

As an extension, propose a hypothetical scenario for students to discuss – for example, 'To reduce global warming the United Nations is proposing banning petrol and diesel cars. Discuss.' Then ask the students to imagine that the number line is like a pair of scales: if you add weight to the right-hand pan (positive numbers) the scales will tilt on one side; if you add weight to the left-hand pan (negative numbers) the scales will tilt the other way. In the process of developing their argument, both positively and negatively weighted ideas will be proposed which means the scales will be constantly adjusting to these inputs.

As a further extension, as students add their pros and cons to the number line they should keep a cumulative (running) total of the value of the statements. It's useful to place a marker on the number line which indicates if the pro or con side of the argument is winning and by how much. The marker should be moved after each new statement and argument is added to the number line, according to the value.

Activity types

Preparation for extended writing tasks – The students are given a proposal such as, 'The Internet is a dangerous place', and then asked to consider questions such as, what justifies this statement, what contradicts this statement and what alternatives should be considered?

Generating discussion points – A similar topic could be used for a live debate, but it's important to get the students to prepare their arguments by gathering evidence (pros and cons) before they stand up and speak.

Forming and debating opinions – A distinction should be made between factual information (for which there is objective evidence) and opinion (an individual's point of view).

Example activities

English – Writing to argue for and against statements.

English – Consequence lines (follow an argument through to investigate some possible scenarios).

Science – Advantages and disadvantages of different types of renewable energy.

Geography – Positive and negative effects of tourism.

Religious education – Discussion of controversial topics such as the legalisation of drugs, abortion, etc.

Mathematical tips

Swapping between the words 'negative' and 'minus' is important so students become proficient in the use of mathematical language.

Variations

Ask students to adopt a specific point of view – possibly one that is opposite to what they initially believe – and argue from that position.

8. Headline Figure

Newspapers and news media cite statistics and percentages on a daily basis (sometimes accurately!). If you scan the media you can use these figures, when appropriate, to get students hooked into the lesson and talking to each other about maths.

Preparation and resources

You will need to find a news headline, newspaper or Internet article relevant to your subject which includes a statistic or numerical fact – it can just be a key fact about an event which includes a number. Project this onto the board. I find it useful to have two headlines ready for the lesson: the first one we complete together as a class, with guidance; the second one is for the students to work on by themselves.

How it works

Selecting a headline figure can be tricky. You want to choose one which could be interpreted in several different ways – for example, you could imitate the old *Guardian* newspaper series called 'Number of the week' or find number-based stories from newspaper headlines such as, 'Penalty points and fines for using a phone while driving will double, to six points and £200'. I often choose one or two different numbers which are percentages, fractions or monetary amounts. When revising information, you could choose a series of figures that students should know about a given case study.

I particularly like to use large numbers such as 8,000,000,000 (8 billion). When I initially present this number to students for interpretation, I intentionally omit the word 'billion' in order to throw them off the scent of the true context and meaning of the number. This also could lead into a brief discussion with the students about the importance of place value and units of measure. Without context a number is often meaningless.

The headline figure doesn't necessarily need to come from a newspaper article; however, it is advantageous to use current information and topics from the media that relate to your subject area.

1. Write a headline figure on the board with initially no units of measure and ask students to discuss what this represents in small groups. For example, '7'.

2. Bring the class back together and discuss their ideas.

3. Next, provide a unit of measure for the figure and see how this changes the students' opinions. For example: 'Magnitude of 7'.

4. Discuss together how their understanding of 'magnitude' changes the meaning of what 7 represents.

5. Finally, give students a context for the figure and lead the discussion into the lesson for the day. For example: 'An earthquake with a magnitude of 7 in Haiti in 2010.'

Activity types

Starter/plenary

Revision

Example activities

Economics – Write the rate of inflation on the board and let the students discuss what this means.

Geography – Write the population of a country on the board and encourage the students to discuss.

History – Use the mortality rate of the Great Plague as a starting figure for discussions.

Mathematical tips

Make sure the students understand the scale of the numbers you are talking about – for example, earthquake magnitudes use a logarithmic scale.

9. Code Breakers

A challenge or puzzle can often make the driest of topics interesting to students. Code Breakers allows students to become totally engaged in a task, absorbing the information as they try to crack the code. This exercise requires a number of mental skills to solve the puzzle.

The intensity of focus I have seen with students engaged in this type of activity is immense. This level of concentration improves their recall of the passage. It seems that they are so involved with decoding the piece of text that they come to know the content inside out.

Preparation and resources

Take a piece of text and translate this into a code for students to solve. This could be anything from a sentence to a paragraph from a source. Here are some simple substitution codes you could try.

Exchange each letter for a number following a pattern:

a	b	c	d	e	f	g	h	i	j	k	l	m	n	o	p	q	r	s	t	u	v	w	x	y	z
1	2	3	4	5	6	7	8	9	10	11	12	13	14	15	16	17	18	19	20	21	22	23	24	25	26

Shift the alphabet along (Caesar cipher):

a	b	c	d	e	f	g	h	i	j	k	l	m	n	o	p	q	r	s	t	u	v	w	x	y	z
z	a	b	c	d	e	f	g	h	i	j	k	l	m	n	o	p	q	r	s	t	u	v	w	x	y

Morse code:

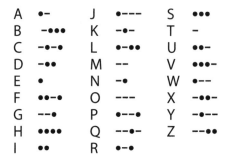

Another method of code cracking is to prepare a series of questions for the students which have one-word answers. The first letter of each correct answer correlates to a different letter in the code-breaking grid. Once all of the questions have been answered the students should have a clear code-broken sentence or phrase.

How it works

1. Provide students with a piece of text which has been written in code.

2. Students then need to use all of their knowledge of a subject or topic to decode the source by looking for word lengths and word structures. They need to decipher the key words as this will help them to decode the full piece of text.

3. It is often easier for students to crack the code if you give them a sheet with the alphabet printed in a grid on it (as above) to help them.

It is at your discretion whether you give them hints about the type of code or leave them to solve it completely by themselves. I suggest building up to the latter once they have become familiar with using and solving different types of codes. Pointers might include clues about sentence structure or some words already translated.

Activity types

Reading sources and texts

Revision/review

Starter/plenary

Example activities

History – When studying the First World War, simulate a war room and deliver a series of Morse-encoded statements including key facts about the conflict for students to decipher.

Business studies – When looking at economies or businesses in different countries, provide students with the sentences using the language of the country being studied.

English – This could be used as a revision strategy to help students memorise key quotes for their final examinations where they aren't allowed the text. As students come to recognise more of the quotation, this will speed up their decryption.

Mathematical tips

Students may find an alphabet wheel useful when solving a Caesar cipher, where the internal ring of letters shifts round.

Further reading

Simon Singh, *The Code Book: The Secret History of Codes and Code-Breaking* (London: Fourth Estate, 1999).

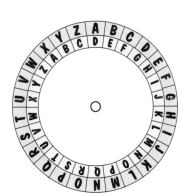

Engagement

10. Twisted Figures

It is often claimed that you can prove anything with statistics. It is true that opinions and decisions can be swayed by presenting data in a certain way. However, because a statistic is an abstraction that condenses an amount of information into a single number, it is easy to see how biases can occur.

Many people who are not familiar with statistical concepts and principles are easily influenced by numbers, for good or bad. Since there are times when the data is deliberately skewed or even misrepresented by government or the media, it is important for students to be aware of potential bias, and to make a point of examining both sides of every story and looking at the underpinning principles on which the conclusions have been based.

However, for Twisted Figures, the students are going to define two opposing points of view through interpreting the same basic data, so that it becomes an exercise in how you state things to emphasise a particular position. This is not about deliberately distorting the data, but about presenting two interpretations which show a complementary bias.

Preparation and resources

You will need to have prepared some facts and figures from headlines or articles that are expressed as fractions, percentages or ratios. These figures should be related to the topic you are studying.

How it works

Twisted Figures asks students to work with the converse of the stated statistic. Having determined one principal argument based on the figures presented, they then have to see things from the opposite point of view and find a way of interpreting the figures so that they now support the counter explanation. Students need to rethink the arguments proposed by the initial supporting data by, for example, calculating new fractions, percentages, decimals and graphs to create a strong and secure opposing argument.

1.   Provide students with newspaper articles or facts and figures on a certain topic – for example: '52.5% of the UK voted to leave the European Union with a majority vote.'

2.   Split the students into for and against the topic to be debated.

3.   Students then need to reinterpret the figures so that they support their side of the argument. To do this they need to be able to rewrite the statements to emphasise a particular point of view. For example, the data might show that '52.5% of the UK voted to leave the European Union with a majority vote'; however, by focusing on the percentage of people who chose to remain rather than leave it is also true to say that '47.5% of the UK voted to stay within the European Union'.

4.   Students should then debate the topic in pairs or as a class, ready for an extended writing activity.

Activity types

Writing to argue

Discussions and debates

Example activities

Media studies – Students are asked to highlight all figures and statistics in a newspaper article. Then they have to annotate their highlights and reverse the statistics – that is, turn the statement around and focus instead on what has *not* been said. This can lead to them writing an article of their own in which they put forward the opposite opinion – turning a negative article into a positive one or vice versa.

Tutor group – This is a good activity to use when looking at attendance with form classes. If the data shows that 'average attendance for students in England is 94.9%', then by focusing on the absence rather than the presence, it is equally true to say that 'students are missing on average 5.1% of the academic year'.

Geography – When studying human geography, provide students with the percentages and figures that result from gentrification in a local area. Split the students into those for and against gentrification. The students then need to develop their arguments for a class debate by selecting the statements that are already relevant for their stance or by twisting the figures and making them useful to their argument – for example, 'Two-thirds of the local population have a job since the opening of the new shopping centre', instead of, 'One-third of local people are still unemployed'.

Mathematical tips

Percentage means 'parts per 100' – that is, the proportions have to add up to 100%.

11. Bargain Words

In any lesson, students will be more motivated to learn if they have had previous success with that subject matter and are familiar with the key concepts and terms. Once they get it, they are more likely to pay more attention and take in new content. In order to increase students' focus on the content, this exercise offers them rewards for noticing when particular words are used. Using monetary values and probability estimates, Bargain Words asks students to predict which words will appear more frequently throughout an activity or lesson. By adding the element of competition, students will be further motivated to become the 'richest' by the end of the activity.

Preparation and resources

You will need to have pre-selected a list of key words, characters or terms from which the students select three. These words should relate to important concepts connected with a set activity, such as reading a passage of text or watching a video clip. When using this activity in a revision lesson, you may want the students to come up with their own key words without any support from you.

You should also provide the students with three monetary amounts that are close in value that they can assign to their three chosen words – for example, £1, £1.50 and £2.

You also have the option to print out the Bargain Words resource sheet (see page 105), or ask the students to draw their own frequency table in their books. This will enable them to keep track by tallying how many times a word appears or is said.

How it works

1. Students need to select three key words, characters or terms that will appear during an activity. They could choose them from a grid of options provided by the teacher or draw up their own list. Students should try to select words that they believe will be used more frequently throughout a video, passage of text or whole-class discussion.

2. Once the students have selected their words they should draw up a table and assign one of the three monetary values provided to each word. The word the students feel will occur most frequently should have the highest value.

3. As each word occurs within the selected activity, the students should keep a tally of how many times they hear or see it.

4. At the end of the activity, the students should calculate the frequency of each word by totalling their tallies. Then, using their selected monetary values and frequency of each word tally, they should calculate how much money each word has earned them and then calculate their overall total. The student with the highest total value of words is the winner.

Who was the richest? Which word appeared most frequently? Why?

Activity types

Watching educational videos – To make sure all students are fully focused and not just dozing during the video, provide them with a challenge which makes them listen to every word said.

Reading texts – In addition to asking students to tally the frequency of their chosen words, you could extend this exercise to make sure they are reading the text by following it up with a series of two or three true or false questions based on their selected key words and the text.

Example activities

Religious education – When watching a video on an important concept, ask students to select key words which they judge will appear in the film.

Computing – When discussing programming, ask students to list key coding phrases as key terms. Which are the most common coding phrases? Is it <\span>? Or could it be <p><\p>?

11. Bargain Words

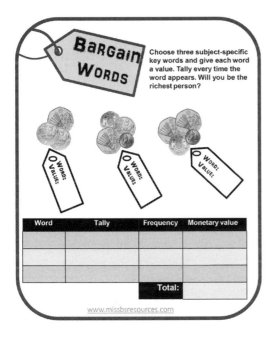

Choose three subject-specific key words and give each word a value. Tally every time the word appears. Will you be the richest person?

Word	Tally	Frequency	Monetary value
		Total:	

www.missbsresources.com

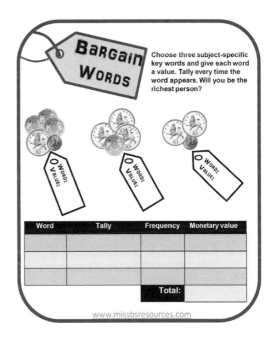

Choose three subject-specific key words and give each word a value. Tally every time the word appears. Will you be the richest person?

Word	Tally	Frequency	Monetary value
		Total:	

www.missbsresources.com

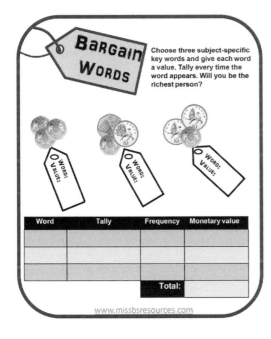

Choose three subject-specific key words and give each word a value. Tally every time the word appears. Will you be the richest person?

Word	Tally	Frequency	Monetary value
		Total:	

www.missbsresources.com

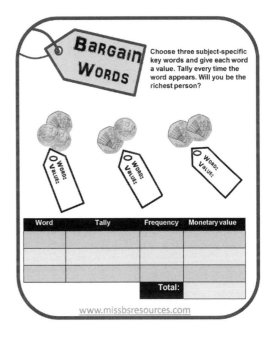

Choose three subject-specific key words and give each word a value. Tally every time the word appears. Will you be the richest person?

Word	Tally	Frequency	Monetary value
		Total:	

www.missbsresources.com

Mathematical tips

Frequency means how often – how many times did the event occur? This is found by totalling the tallies.

When students are writing monetary amounts, make sure they are written correctly – for example, £2.40 and not £2.4 or £2.40p.

Variations

According to the ability level of your classes, vary the monetary values assigned to the words.

Allow each student to come up with their own three words when visiting a topic for a second time to encourage them to recall what they learned in the previous lesson.

12. Put the Fire Out

This exercise is similar to the Impact Line (Resource 7) but it is a shorter activity and can be tailored for higher ability students to involve the addition of non-integer values. However, this activity requires students to think using different units of measurement.

Put the Fire Out is a great way of getting a class to think outside the box, as students are asked to come up with arguments that disprove common myths and misconceptions, such as: everyone needs to drink 2 litres of water a day, the Great Wall of China is visible from space and we only use 10% of our brain. Many students will struggle to explain why something is right or wrong in words or form valid arguments to illustrate their points.

This exercise uses the metaphor of filling a tank with value-assigned points which refute the initial misconception. Put the Fire Out focuses on students' concept of number and measure and asks them to build strong arguments for or against a topic/statement.

Preparation and resources

You will need to either project the Put the Fire Out worksheet (see page 108) onto the board or provide each student with a copy of the worksheet. This can be laminated for reuse.

How it works

1. Students work in small groups. At the beginning of the exercise they are given a short period (e.g. five or ten minutes) to think about the topic and make notes.

2. Explain to students that they need to fill the firefighters' tank with water in order to put out the fire that you have created by deliberately stating a misconception or red herring or by playing devil's advocate.

3. Students should give simple reasons for why a statement isn't true; each reason gains them 100 litres of water for their tank.

4. Students then need to justify and explain the reasons behind their arguments – evidence should be provided where possible with quotes, facts and figures. Each justification and explanation gains the students 300 litres of water for their tank.

5. Students need 1.5 kilolitres (1,500 litres) of water for the tank to be full. Make a point of getting the students to work out the conversion. If working with higher ability students, it may be appropriate to get them to work in kilolitres, which means they will be working with the addition and subtraction of decimal numbers.

6. Once the tank is full, each student should have a completed argument to present their case or to be used as the beginning of an essay.

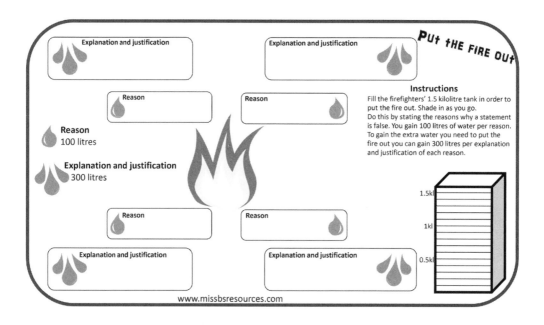

Activity types

Planning extended writing

Creating arguments/writing to argue

Discussions and debates

Example activities

Food technology – 'All processed foods are bad for people's health.'

Music – 'Twenty-first century music is just a load of noise with no melodies.'

PE – 'Athletes should be allowed to take some performance-enhancing drugs.'

Teacher notes

This exercise is useful in getting students to think through possible arguments to support a case. Keep the reasons coming. It should be a fairly quick process as it is mainly about gathering ideas. Don't spend too much time debating the issue. Forming the ideas into a coherent essay will require students to think further and deeper on the argument they are proposing.

Mathematical tips

One kilolitre is equivalent to 1,000 litres; this means that 0.5 kilolitres is equal to 500 litres.

Variations

You may want students to create their own questions or task grids when a topic is completed, which they can use later to quiz themselves for revision purposes.

13. Fuel Fill Race

Fuel Fill Race is a very simple idea based on the principle of knowledge is power. In order to succeed, students need to fill the fuel tank of a car. This involves both money and measure, and a choice of tactics. Students might answer too many easy questions for less money and risk not being able to fill their tank, or they might attempt to answer difficult questions, make mistakes and then find they haven't got enough money left to fill their tank. They have to think in depth about strategy, pace and risk. This is good practice for future exam success, as they will need to gain the highest score possible by choosing the best questions to answer, so they're not throwing away marks or wasting time.

Preparation and resources

You will need a set of 20 questions on a topic ranging in difficulty. The value of each question is related to the cost and quantity of fuel.

Students are given the pre-constructed 4 × 5 grid template (see page 111, perhaps with each square numbered) and the questions are projected onto the board. Alternatively, print the questions on a separate sheet and project the price/quantity chart onto the board. As the questions are answered, students cross out the relevant squares and colour in their tanks.

Another option is to give students colour-coded question stacks, but this takes a lot more preparation time.

How it works

1. In pairs, students are given a fixed amount of money (£60) to spend on questions.

2. The aim is to answer questions which will generate a stated amount of fuel. To win the competition they must reach a total of 40 litres.

3. Students should peer assess this activity. If students answer the questions correctly they get that amount of fuel. The harder the questions, the

greater the value of the square. By answering higher value questions correctly, they will be able to fill the tank quicker.

4. If students answer a question incorrectly they still have to pay the cost of
   the square, but they don't receive the fuel. Once a square has been chosen,
   it can't be chosen again.

5. Throughout the activity, students should be noting how much money they
   have left and how much of their tank still needs to be filled.

6. Make it clear that the total amount of fuel has to be exactly 40 litres. Students aren't allowed to over- or under-fill the tank. This means they have
   to think more strategically about which squares to select.

FUEL FILL RACE

Your car needs 40 litres of fuel to run. You have a budget of £60 to fill your tank. Choose squares around the grid – think strategically because whether you answer a question right or wrong you have to pay the money of the square. Answer too many questions incorrectly and you won't have enough money to fill the tank.

Fill your tank before your opponent to win.

£5	£10	£7.50	£7.50
4l	12l	7l	7l
£7.50	£5	£2.50	£5
7l	4l	2l	4l
£10	£1.50	£5	£2.50
14l	1l	4l	2l
£1.50	£2.50	£5	£5
1l	2l	4l	4l
£5	£5	£12.50	£10
4l	4l	15l	14l

www.missbsresources.com

Activity types

Competition

Group work

Starter

Plenary

Example activities

Science – When reviewing a topic area at the end of a module.

Geography – To build up an answer to a GCSE examination question by chunking the important facts and points into a grid.

Mathematical tips

Look at number bonds: £2.50 + £2.50 = £5 and £7.50 + £2.50 = £10.

Make sure students are writing monetary amounts correctly. For example £2.5 is incorrect; it must be written as £2.50.

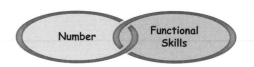
14. Shopping Spree

Managing money is a life skill, so it's vital that students think in terms of budgeting and are able to balance their income with outgoings and avoid running up huge debts. Money doesn't grow on trees so we always need to stick to a budget. Can your students choose wisely given the prompts they will receive?

Preparation and resources

Prepare an open-ended task for students to complete, such as write a poem which relates to gender inequality or design a wildlife reserve to protect endangered species.

Create a series of envelopes containing different prompts, hints and tips relating to the task. For example, for the poem you might include an example poem or a list of stylistic literary devices such as alliteration. For designing a wildlife reserve, you might have a prompt to get students to include grid references or key facts about endangered species and their care.

Price up each envelope, calculating the value according to the perceived help factor of the prompts inside. The better the prompt, the more expensive the envelope.

How it works

1. Provide students with a set amount of fake money – £5 to £10 is a good starting amount.

2. Students then attempt to complete the task. If they get stuck they can purchase a prompt.

3. They need to make the decision whether to spend more money on one envelope or get two envelopes of a lesser value but with more prompts. With no clue as to what is inside each envelope, students need to carefully balance their cash with their need for more information.

4. The prompts then form the foundations of support for the task.

Activity types

Creative writing

Extended writing

Group work

Example activities

English – Provide students with prompts for a story. This could be a series of adjectives, literary device prompts or writing style hints.

History – Provide key dates, information about historical figures or important events.

Mathematical tips

See the functional skills support mat (Figure 4.9) for support when calculating with money.

15. Battle Words

In the 20th century, before the arrival of satnavs and GPS, being able to read a map was an essential part of finding your way around. However, given the digital assistance now available, that art is being lost. Battle Words is about the basics of map-reading – for those times when there is no signal or the battery has run down. I know from personal experience that total reliance on a satnav can lead to unfortunate consequences – I have become lost because my car's GPS decided to direct me along a rough track which led to the middle of a field!

Students need to be able to read and interpret maps in everyday life. For the newcomer, just finding their way around a school building can be a daunting task. It can be made easier if the new student has a plan of the layout of classrooms and so on, but they also need to be familiar with the conventions of maps and know how to extract the information they need.

One of the core skills of map-reading is being able to state a position on a map using grid coordinates. The idea of Battle Words is that students apply their knowledge of coordinates to remember and define key words. The exercise is based on the traditional game of Battleships, using key words to represent naval vessels.

Battle Words is a competitive game with bonus points awarded for guessing the key words before knowing all the letters and for being able to accurately define the terms. This activity would be great as a recap starter, plenary or revision tool.

Preparation and resources

You will need to provide students with a Battleships-style grid. These can be pre-printed and laminated for reuse. Students should also have access to their books and perhaps also a glossary.

How it works

There are two levels of play for Battle Words. The basic level uses a one-quadrant grid and the advanced level uses a four-quadrant grid and is suitable for higher ability students.

1.   Set up. Students select five key words. Each word is given a point score from 1 to 5. The word they believe will be the hardest for their opponent to guess is 5 points, the easiest 1 point.

2.   Students place these words on their grid. A word can run either horizontally or vertically but not diagonally. Letters should be adjoining; L-shapes are not permitted.

3.   The play. Once students have entered their five words on the grid, they take it in turns to guess what their opponent's words are by selecting a set of coordinates and asking for information about the content of that particular square.

4.   If student A scores a hit then student B tells student A the letter. Student A then writes this letter in the same grid square on their sheet. In this way students can keep track of their hits and misses.

5.   As the students fill in the squares, they can at any time guess the word. They get a bonus point if they guess the word correctly before it has been completed.

6.   Whenever a key word is revealed (i.e. a ship is sunk), the student gains the points value of that ship. In addition, if the student defines the word correctly they get a bonus point.

Scoring system

As the game progresses, students tally their scores in the table provided.

   1 point – for a hit

   Ship score of 1 to 5 – for sinking a ship

   1 point – bonus point for guessing the word early

   1 point – bonus point for defining the word correctly

Activity types

Reading texts

Recap

Starter

Plenary

Revision

Example activities

History – Provide a source text and ask students to pull out key terms on their second read-through. When students have guessed their opponent's key words correctly, they then need to provide the context of the key term or person for a bonus point.

Science – As a way of reviewing the content of a previous lesson. For example, ask students to create a list of key words associated with photosynthesis.

Mathematical tips

Coordinates are written in the form (x, y) never (y, x). If students are struggling to plot or read coordinates, remind them of the mnemonic, along the corridor and up the stairs.

Another way to remember this is:

Horizontal is across – the x axis (remember this as 'x is a cross')

Vertical is up and down – the y axis (remember this as 'y to the sky')

In the Battle Words resources, the x and y axes are shown and labelled as you would with a standard coordinate or number line grid.

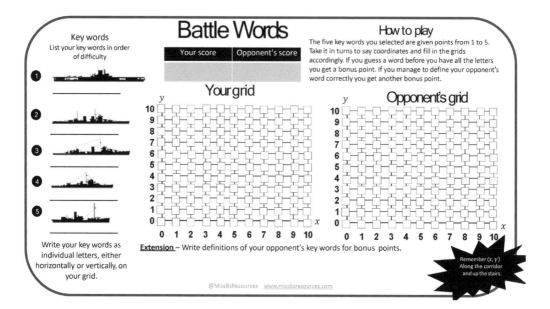

Battle Words

Key words

List your key words in order of difficulty

①
②
③
④
⑤

Write your key words as individual letters, either horizontally or vertically, on your grid.

Your score	Opponent's score

Your grid

Opponent's grid

How to play

The five key words you selected are given points from 1 to 5. Take it in turns to say coordinates and fill in the grids accordingly. If you guess a word before you have all the letters you get a bonus point. If you manage to define your opponent's word correctly you get another bonus point.

Extension – Write definitions of your opponent's key words for bonus points.

Remember (x, y) Along the corridor and up the stairs.

@MissBsResources   www.missbsresources.com

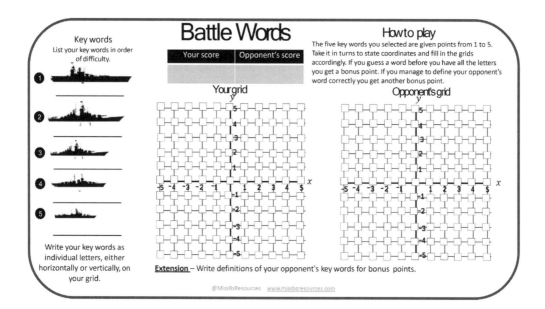

Battle Words

Key words

List your key words in order of difficulty.

①
②
③
④
⑤

Write your key words as individual letters, either horizontally or vertically, on your grid.

Your score	Opponent's score

Your grid

Opponent's grid

How to play

The five key words you selected are given points from 1 to 5. Take it in turns to state coordinates and fill in the grids accordingly. If you guess a word before you have all the letters you get a bonus point. If you manage to define your opponent's word correctly you get another bonus point.

Extension – Write definitions of your opponent's key words for bonus points.

@MissBsResources   www.missbsresources.com

16. Netting Questions

Some students have difficulty in drawing nets of shapes. However, they can be introduced to the art of folding a flat sheet into a three-dimensional cube using the pattern below. The net doesn't have to be limited to a cube; other polyhedra such as octahedra or icosahedra can be constructed using a similar procedure. The finished cube or polyhedron can then be used as a die or for writing prompts.

Preparation and resources

Provide students with a piece of card to draw their own net or print the net of a shape onto a piece of card for them to cut out. You will also need some glue or sticky tape.

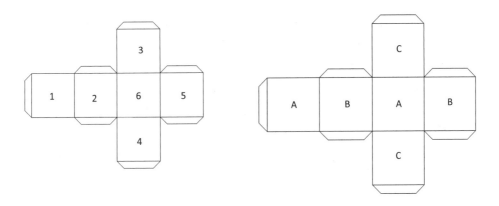

Net for a cube

Remember to include tabs (as shown) when designing a net so that you can glue the faces together.

Other shapes you may wish to consider are the tetrahedron, octahedron and icosahedron:

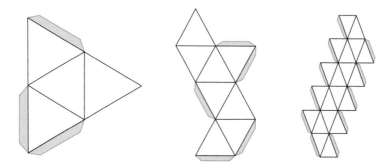

How it works

1. Either hand out a pre-printed net or ask students to draw a net of a 3D shape. The easiest to construct is the cube.

2. Students should then write questions or prompts on each face.

3. Once the net has been cut out and glued together, students can use it to provide a random input – either as something to discuss or learn or as a prompt for questions.

4. Certain instructions or questions can be repeated on more than one face so as to increase the probability of that result occurring.

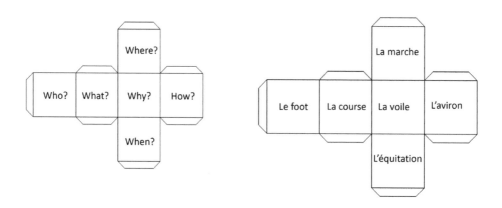

Activity types

Revision/review

Creative writing

Studying case studies

Learning vocabulary

Example activities

History – When reviewing key events, use a who, what, when, why, where and how die to draw out the facts from students. They can work in pairs and ask each other questions based on the outcome of the roll.

Geography – When revising for an exam, place the titles of key topics or case studies on a cube or octahedral die. Once the die is rolled, students need to write down as many facts as they can about the topic.

MFL – Write key vocabulary on the current topic on the die and ask a student to translate the term that appears on the uppermost face. A nice idea for extending this resource is to write translated and untranslated key words on opposite faces of the die. Before they assemble the cube, students would need to identify which faces will be opposite each other and then write the translated and untranslated words in the correct places.

Mathematical tips

Fact: The opposite faces of a standard die total 7.

Marking and Reflection

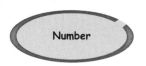

17. Thermometer of Understanding

This exercise helps to promote students' level of confidence in understanding a particular topic by honestly assessing their existing knowledge so they realise how much further they need to apply themselves. Students may want to work on this exercise individually if they do not wish to make their self-assessments public knowledge. If they are making their assessments public, then it may be useful for them to get feedback on their level from their classmates in case they are over- or undervaluing their expertise.

It is a common practice in many classrooms to ask students to RAG-rate their understanding or feelings either during a presentation or at the end of a lesson or topic. If you are not familiar with this way of checking students' understanding, you simply get them to choose the colour from the traffic light structure (red, amber or green) that best represents how they feel or how confident they are with their understanding at this moment.

Although the upper and lower quartiles of your students are usually honest, the middle 50% often either undersell themselves if they are on the amber/ green borderline or oversell themselves at the amber/red borderline. Students also tend to check what everyone else is doing before revealing their choice as they don't want to look silly or be the odd one out. It's also likely that students who are poor at making judgements will find it challenging to estimate their true level of ability or will have little confidence in themselves. You also need to take into account an effect – which is particularly prevalent within higher sets – when students realise there is even more to learn about a topic than they originally thought, so they don't RAG on their knowledge so far but are instead comparing their perceived amount of knowledge with what they will eventually need to know.

In other words, there are a number of ways in which the RAG system can lead to inaccurate results. To partially get around this, I propose using the Thermometer of Understanding. Instead of the three colour categories, students are asked to estimate where they are on a continuum – in this case, a number line which has both positive and negative values, as with a standard measurement of temperature. The number line allows teachers to identify the indecisive red/amber and amber/green students easily, thus making it easier to assess the students' understanding at a glance.

The aim is to increase the temperature of students' confidence on a topic or concept over time. However, this is unlikely to be a smooth or gradual improvement; their journey through the topic will include highs and lows, with occasional plateaus.

Preparation and resources

You will need to project the Thermometer of Understanding onto the board or you could display posters on the walls instead. Some teachers find it useful to print out mini-thermometers for the students to colour in; others get them to note the temperature in their books. (Remind them to write °C afterwards.)

How it works

1. At the start or end of a lesson, topic or concept allow students time to reflect on their current 'temperature' and why they think this is the case. Students should talk about whether their temperature has increased or decreased, and what factors may be responsible for this journey.

2. Once the students have responded using the number line, you can clearly see where intervention might be needed in the form of either confidence boosting, support or stretch activities.

3. Each student can keep a record of their day-to-day temperature for each topic or subject. From this, they will be able to plot their learning over time. This can often be interesting, both to themselves as well as to you, as there may be a pattern to when the highs and lows happen – for example, period 1 on a Monday or period 5 on a Friday. This allows students to identify visual trends in their own behaviours and potentially self-correct or seek support.

This idea came from Fiona Ritson and has been used with permission.

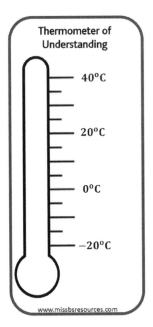

Activity types

Reflection on work, concept or topic

Marking and feedback

Example activities

English – Assessing confidence in using grammar, spelling and punctuation.

French – Assessing confidence in using different tenses.

Mathematical tips

Discuss the concept of the number line with students and clarify that -5 is less than -1.

Teacher notes

Self-confidence often drops when students discover that there is more and more they don't know! As beginners, they start with the feeling that they are succeeding, but when they realise how much there is still to learn they can start to feel overwhelmed. This should pass once they appreciate that everyone goes through this phase. To counter this, it can be a good idea to get them to reflect on how far they have come already.

18. The Real Value

Every class seems to have students who mess up their books or somehow manage to make a pen break into a thousand pieces in front of your eyes. Do these students truly value what they have? This exercise asks them to put a value on their possessions and their work. As with any job, they should know that their work may be evaluated by their peers in due course.

Preparation and resources

Little preparation is needed for this activity. You can either print out slips for the students to write on or they can complete this activity on scrap paper that will be collected in at the end. I have found that it is important to give students fair warning when you first introduce this activity, so I often tell them at the beginning of the week that we will be doing this exercise at the end of the week.

How it works

1. Students are asked to assign a meaningful value to their exercise books, thus making them into a prized item.

2. At the end of each week, I choose five books at random and place them on a desk at the front of the room. The books are opened at a page of work of which the students are proud. It is important that they are anonymised. I then ask the other students to rate the books for effort, presentation and whether the content makes sense.

3. Each student rates the book as they would an online product using a five star system.

4. They then assess the value of the book as if it were a revision guide. How much would they pay for the book between £0 and £10?

5. Students place their ratings in a plastic wallet which is put inside the book for the individual student to read. This student should be given time in the next lesson to look at the feedback and then work out their averages and the range of the ratings and 'price' by calculating the mean, median, mode

and range. Hopefully, their individual average scores will improve over time and the range of their results will decrease, meaning peers are being more consistent in the marks they are awarding to each student.

6. This activity also gives students a chance to gather ideas from each other on presentation styles. They gain an understanding of what useful notes look like and what isn't useful.

The real value

Effort ☆☆☆☆☆

Presentation ☆☆☆☆☆

Value: £_____

Activity types

Marking and feedback

Example activities

Design and technology – This could be carried out at different stages of a product development task to help students keep on task and up to speed and gain plenty of ideas.

English – Discussion about presentation, how people deal with mistakes and clearly notating paragraphs.

Mathematical tips

To calculate the *mean*, add up all the numbers and divide by how many there are.

To calculate the *median*, place the numbers in ascending order and then identify the mid-point number. For an even amount of numbers the median is halfway between the two middle values.

To calculate the *mode*, place the numbers in ascending order and see which item or number appears most often (there can be more than one mode). If no items or numbers appear more than any other then there is no mode.

To calculate the *range* (to assess the consistency of the class's opinion of the work), subtract the smallest value from the largest value.

Variations

An extension to this task would be for students to leave a written review of the work while peer assessing.

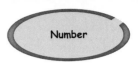

19. Maths Marking

Marking takes a significant chunk of time to complete, so having a way of saving yourself from having to write out the same statement or comment multiple times can speed up the process. This exercise asks students to use conventional maths symbols as a kind of marking shorthand. Using the symbols is an effective way of exposing students to mathematical notation.

Preparation and resources

Make sure that copies of the marking guide are clearly visible in the room, either on displays or projected onto the board. Alternatively, the students could stick a small version of the guide into their books.

How it works

Marking books can often get tiresome with some teachers having to use different coloured pens to mark different ideas. To speed up marking, acquaint yourself with these common mathematical symbols and meanings, so instead of writing words or sentences you use these mathematical symbols to give quality feedback at a faster pace. This is by no means a replacement for all formal written feedback; however, it is a useful tool in your teaching toolbox.

This process will not only speed up your marking, it will also allow students to see these important mathematical symbols in relevant contexts to help secure their meaning. It is important that students have solid foundations for mathematical concepts, particularly the more obscure symbols such as sigma, which some of them may not come across until their later years in school.

This strategy can also speed up the marking dialogue between students and teachers, encouraging students to amend their own work. It can also be used as a good peer-assessment tool in lessons.

Marking Guide

$+$	Positive	This is good
$-$	Negative	This isn't so good
$\therefore$	Therefore	Explain effect
$\neq$	Not equal	This doesn't relate to the question or make sense
$\cup$	Union	Join these together
$<$	Less than	Less about this point
$>$	More than	More about this point
Σ	Sigma	Sum it up
∞	Infinity	Waffle

www.missbsresources.com

Activity types

Feedback and marking

Peer and self-assessment

Example activities

English – Introduce this method to help students when they are peer assessing extended writing tasks.

Geography – When marking a student's first draft of an extended writing assignment, use these symbols to highlight the good and bad bits as well as any corrections or additions the student needs to make.

20. Graph It

Graph It is a simple way for a student to track their level of understanding and progress of a topic over time. By using this as a basis for a learning journal, you can easily create a dialogue between yourself and the student. It is important for students to take ownership of their own learning and also to be self-reflective and aware of the journey they are on – without under- or overselling themselves. When preparing students for life beyond school, it is vital that they are able to identify their strengths and weaknesses in order to know how to improve.

The learning journal provides students with an instant go-to place to find out what they need to work on to progress. It also gives them a confidence boost as it shows how they have struggled and overcome various uphill battles.

Preparation and resources

Students will need a piece of graph paper or squared paper and to be given time to draw the axes of the graph: time in weeks along the x axis and self-assessment score on the y axis (from 0% to 100%).

How it works

1. Students should create a visual diary in their book of their understanding of a topic or concept over time in the form of a graph.

2. You can decide how often you want the students to plot their understanding – they could do it for every lesson, every week or at the beginning, middle and end of a topic. Students should plot their understanding with their work in mind: did they get 100% on the examination questions? Did they make some silly mistakes that effected their end result? Or do they have a big conceptual misunderstanding that needs sorting out?

3. The second, but no less important, part of this process is getting the students to annotate the graph and note down their reasoning for plotting particular values. This is a useful way of highlighting areas of strength and weakness. In addition, you may wish to ask them to comment on how

they think they can help themselves or what they might require from you in order to improve. You can then use the students' comments to open up a dialogue with them, as well as provide students with strategies and words of encouragement to help improve their confidence and understanding where appropriate. This will help to make each student feel valued.

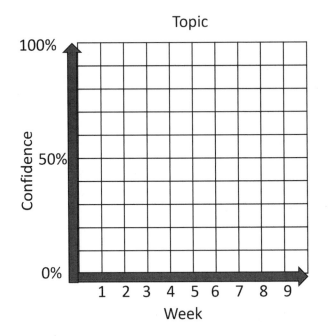

Activity types

Self-evaluation

Marking and feedback

Example activities

English – When studying texts, ask students to plot their understanding of the relationships between the characters. Do they understand the dynamic between them? Are they confused about what is going on? Have they reached a new understanding of an earlier situation now that something else has happened in the book? How do they feel about revising an idea that they initially found difficult?

A variation on this is for students to plot the relationship between two characters on a graph over time – for example, in *Romeo and Juliet*, this could show the highs and lows of their love affair and plot the key events. Provide students with a pre-drawn graph and ask them to determine and then annotate on the graph what the key events were in the play which brought about the changes in the graph.

History – Provide students with a series of examination questions and ask them to plot their confidence level on their ability to answer the questions.

Mathematical tips

Students should always use an accurate scale on each axis and label the axes. Refer to the graphs and statistics support mat (Figure 4.4) for a checklist on drawing a graph.

Variations

A nice variation of this activity is to run two line graphs on the same chart. One should be self-assessment on a student's perceived understanding and the other should be peer assessment on what their book work suggests about their perceived understanding. This will help to regulate those students who either overrate or undersell themselves.

21. Results

(Variant of Resource 20 – Graph It)

Getting feedback is a vital part of learning. Students not only need to be able to track their own results and understanding, but also know how to improve. Boosting independence by breaking down results helps students to see clearly where they may be going wrong. They need to be aware of both their strengths and their weaknesses. This activity helps to track the progress of students over time.

It is important that all students seek to improve their understanding of a subject to enable the improvement of scores/percentages, and if these are improving then the grade should follow. For many students, increasing their grade can seem like an impossible challenge, but by focusing on improving their score by one to five marks at a time, these marginal gains will ultimately impact on the final grade. Students' confidence will also improve through regularly celebrating the small advances they make.

Preparation and resources

As the teacher, you will need to decide which tests, essays or classwork you want to assess so that you can give the students a baseline score. Practice examination questions are a great option for both teacher and students. They not only enable students to become familiar with what the exam board is looking for, but they also provide an opportunity to build in peer feedback and give students a chance to read other students' work and encounter different approaches to the task. This experience is often invaluable.

How it works

1. Students should either get back peer-marked or teacher-marked work, along with an examination score which correlates with a mark scheme. This should state the maximum number of marks available for each question/task.

2. The students then calculate their score as a fraction of the total possible score and then convert this into a percentage.

3. They should then add their latest result to a bar chart or line graph of previous results on topics of a similar nature. I often ask students to annotate their graph and comment on what has improved since last time and what their next steps will be. This should turn the discussion from the actual score into a more nuanced skills-based discussion.

4. From time to time you should ask students to reattempt some questions so they have a chance to improve their score. For some subjects, this might entail students first looking at the exam criteria and then seeing how they could boost their original answers.

Activity types

Marking and feedback

Exam analysis

Example activities

Science – When trying to help students get used to open-ended compare and contrast questions. They are able to track their progress over time as these questions are typically awarded a similar weighted score and will be comparable.

English – Create a dual bar graph of the different papers and elements of their GCSE exam. This will allow students to identify visually which aspects need more of their attention.

Mathematical tips

See the number support mat (Figure 4.13) for percentage guidance and see the graphs and statistics support mat (Figure 4.4) for guidance on graph drawing.

Organisation and Presentation

22. Up to Date

The date is a topic often taken for granted. However, students need to know the various ways of writing the date – in full, in shorthand or as computer-sortable. They also need to know that other countries, such as the United States, do things differently.

There are many students who are not sure of the order of the months in a calendar year and subsequently struggle to understand the numbers shown in a shorthand date. This piece of general knowledge is a requirement throughout adult life – for example, when going to job interviews, catching holiday flights, remembering birthdays and scheduling events.

Preparation and resources

You will need a place in your classroom where the date can be clearly displayed.

How it works

There are several different ways to use this idea. Below are some suggestions on how you could use this activity in your classroom:

- Make writing the current date part of the everyday ritual of the classroom. Write out the date in longhand – for example, 2nd November 2017 instead of 2 Nov 17 or 02/11/17 – to distinguish it from being viewed as just a sequential list of numbers.

- Displaying the date in its full form is best practice, but it doesn't mean the students must copy down the date in this way. It could be argued that transferring the full date to shorthand is a good way of consolidating their understanding of the order of the months, and vice versa. It comes down to personal preference – but only once the students are familiar with the full date.

- It is a good idea to share facts about the date that relate to your subject – for example, 23rd April is the anniversary of William Shakespeare's death in 1616, and 20th July is the anniversary of the day that American astronauts first landed on the moon in 1969. This gets students discussing dates and, naturally, trying to calculate how long ago the event was.

Activity types

Organisation of work

Recital of key event dates

Example activities

MFL – You could write the date using the foreign language the students are learning and ask them to translate the date into English, or vice versa. Remember that capitalisation may not be the same as in English.

History – Defining and recording the dates of significant key events.

Mathematical tips

1 – January (31 days)
2 – February (28 days/29 days in a leap year)
3 – March (31 days)
4 – April (30 days)
5 – May (31 days)
6 – June (30 days)
7 – July (31 days)
8 – August (31 days)
9 – September (30 days)
10 – October (31 days)
11 – November (30 days)
12 – December (31 days)

There is a well-known rhyme for remembering the number of days in each month:

Thirty days hath September,

April, June, and November;

All the rest have thirty-one,

Excepting February alone,

And that has twenty-eight days clear

And twenty-nine in each leap year.

Teacher notes

In the United States, the date is ordered differently, with the month first followed by the date and year – for example, 10/5/2016 is 5th October in the US and 10th May in the UK. In the US, 9/11 refers to 11th September; in the UK we would write 11/9/2001.

Variations

For EBacc students and those studying languages, it might be useful to display the date in other languages for the students to translate back into English.

Discuss adding and subtracting dates, and how difficult this is using conventional systems. This is why many calendars and diaries include the day number from 1 to 365 to make the calculations easier.

You can also introduce computer-sortable dates – for example, 171102 (for 2nd November 2017) – which makes sorting files into date order easy.

23. Diagram Scales

We frequently ask students who are collecting or sorting data in a classroom activity to draw a diagram or construct a table that will organise what they have found. This is a relatively simple task, and one that can be significantly improved if they present their findings in a neatly drawn table or diagram. This exercise gets them to draw tables and boxes using a straight-edge or ruler.

Preparation and resources

You will need to have a model example of the table you wish to be drawn on the whiteboard, including precise measurements. All students will need a ruler.

How it works

1. When asking students to copy or draw a table or diagram, it is a good idea to ask them to sketch the desired outcome in rough first in order to know how best to scale the diagram on the page. Once they know how many rows or columns are needed, they can become more specific about the actual dimensions of the chart or diagram.

2. It is important that the students use a ruler and pencil where appropriate (not their lunch card and pen!). Generally speaking, column widths should be adjusted to the amount of information required (e.g. words or figures). If all the information is of the same kind (e.g. figures) then the columns should be of equal width.

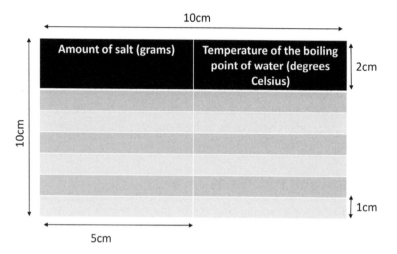

Activity types

Drawing and labelling tables and diagrams

Example activities

Art – When drawing scale drawings, discuss the scale factor of enlargement.

Design and technology – When asking students to accurately draw and label net diagrams for packaging, provide them with the dimensions of the shapes.

Science – When asking students to draw a results table for an experiment, provide them with the measurements for each row and column to help them make a suitably sized table for their results.

Mathematical tips

The scale factor of enlargement is the number of times bigger a shape is in terms of length, area or volume. For example, if you want students to draw two shapes, with one bigger than the other, you could ask them to use a scale factor of 2 which means they would double the length of each side.

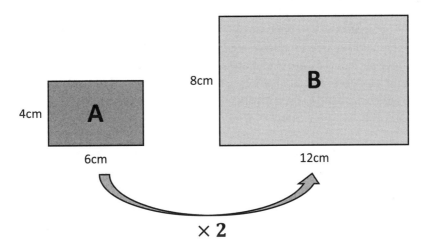

If you want to apply a scaling factor to an area or volume, then you need to know how much larger the resulting area or volume should be. If you want the shape to have dimensions twice the original size, then the new area will be 2 × 2 times (4 times) as big; if you are doubling the dimensions of a three-dimensional rectangular object such as a box, then the new volume will be 2 × 2 × 2 times (8 times) the original volume.

24. Venn Diagrams

Categorising and creating order from lots of information can often be quite difficult for students. One useful aid is the Venn diagram. This is a way of depicting sets or categories which have some unique features and some other features in common. Venn diagrams enable students to visually interpret the information they have sorted. They are then better able to put forward valid arguments when asked to compare and contrast it – for example, the characteristics of a hero and a villain. By listing some distinctive features for both heroes and villains, the students will find that many of them will be opposite characteristics; however, there will also be some characteristics which apply to both heroes and villains. Students need to be able to discuss both the similarities and the differences.

Preparation and resources

It is important to always start with the problem that you wish to solve – for example, what characteristics distinguish category X from category Y? Are there any similarities between the sorting categories? Do some of the characteristics fall into both categories?

You can prepare the categories/sets that you want students to consider and then place key quotes, facts or characters into each category. Alternatively, the students can choose their own categories/sets to organise the key quotes, facts or characters that you provide for them.

How it works

1. Students should classify key words, characters or statements within the correct set (circle) according to your specified criteria. Where a particular piece of data qualifies for two or three sets, it should be placed in the appropriate overlapping section of the Venn diagram.

2. If a piece of information fits into neither set (circle), this should be placed outside the Venn diagram but inside the universal set (the box).

An example is: how would you categorise the following Shakespeare plays into the genres of histories and comedies: *Cymbeline*; *Henry IV, Part 1*; *Henry IV, Part 2*; *Hamlet*; *Henry V*; *The Merry Wives of Windsor*; *A Midsummer Night's Dream*; *Richard II*; *Romeo and Juliet*; *The Tempest*; and *Twelfth Night*?

Shakespeare's Plays

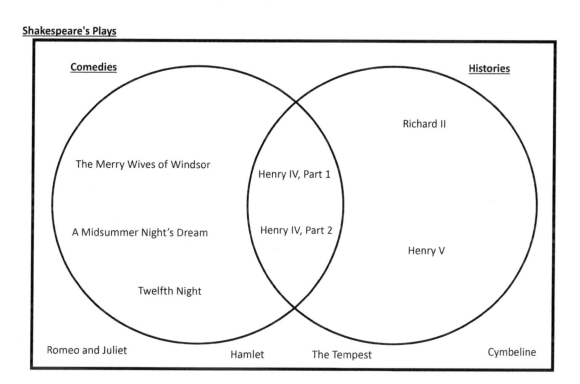

Activity types

Group work

Discussion

Categorising

Comparing and contrasting

Example activities

English – Compare texts or character qualities using adjectives in preparation for an extended writing task. For example, students have to find quotations from a book to support arguments for and against themes such as power, love, money, appearance and reality, fate and free will, heroism and friendship.

Geography – In groups, ask students to place statements into a Venn diagram for positives and negatives or advantages and disadvantages of developing certain industries or regional transport links.

Music – When classifying the common traits of musical genres, ask students to first decide what to label the categories/sets and then where to place them in a Venn diagram.

Mathematical tips

Each circle represents a set or category of information. The intersecting overlap of the circles (sets) represents certain qualities or attributes that belong to both sets. This is the 'and' rule in probability – the overlapping section contains attributes that are in set A and set B (A ∩ B).

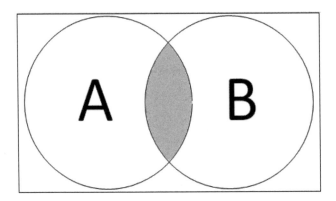

The information in both sets when combined together is called the union. This is the 'or' rule in probability: it applies when an attribute is in set A, or set B, or both A and B (A ∪ B).

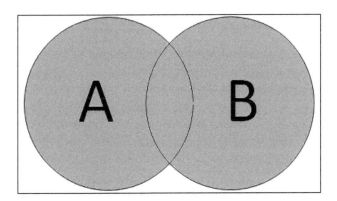

Although it is possible to have more than three sets overlapping in a Venn diagram, in practical terms the maximum number of sets you can show easily is three.

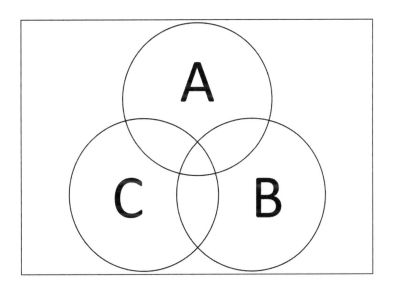

25. Two-Way Tables

This activity has similarities to Venn Diagrams (Resource 24) as it is about the process of sorting and categorising information. However, Venn diagrams and two-way tables are used in different scenarios – for example, in everyday life, two-way tables often appear in the form of a timetable.

This exercise asks students to list and categorise data in order to make connections and spot patterns. Two-way tables play an important part in the development of a student's ability to compare and contrast information – for example, they are a simple visual way to organise revision notes. Moreover, this skill will later help them to make informed decisions when comparing the benefits of different smartphones or choosing where to go to university.

Preparation and resources

You will need to provide students with the two domains of knowledge that they will be comparing, together with the subheadings which identify relevant qualities or categories.

How it works

The students construct a two-way table for organising information into two contrasting categories – for example, wildlife (birds/mammals) and endangered (yes/no). Students enter the terms into the relevant box in the table making sure that each item conforms to both the x and y dimensions.

	Swim	Can't swim
Legs	Sea turtles Humans Ducks Chimpanzees Dogs Shrimp	Spiders
No legs	Jellyfish Whales Sharks Fish	Earthworms

A two-way table can also be an extremely useful data collection tool, often known as a data collection sheet. They are often used to collect a large volume of data about a specific topic – for example, what colour hair do people have?

Hair colour	Tally	Frequency
Brown	III	3
Blonde	IIII	4
Black	IIII II	7
Other	II	2

Activity types

Categorising

Representing information

Example activities

History – Comparisons between the First and Second World Wars.

Science – Comparisons between vertebrates and invertebrates.

26. Going with the Flow

Some students (and adults) find that organising complex information into a logical order can be challenging. However, the process can be made easier by showing the various steps to be taken graphically using a flow chart. The first part of the task is to identify the essential stages of the process – the decisions that need to be made along the way – and then to arrange them in a logical sequence of steps that illustrate how to carry out the task in real life.

Preparation and resources

You will need to guide students through an example flow chart to show them how the concept works. This should cover the basic conventions of a flow chart so they know what the different shapes mean:

- Key stages of a process: action = rectangle

- Key decisions: yes/no or true/false questions = rhombus

- Each answer to a question, yes or no, should lead back to a key stage = rectangle

There are four options when running this activity which require different resources:

1. The simplest way for students to create their own flow chart is to draw it on a piece of paper using a pencil and ruler (they may want to make rectangle and rhombus templates for drawing the boxes). However, you could provide them with ready-made colour-coded sticky notes or cut-out shapes (e.g. all rhombuses are purple and rectangles are yellow). Students then write their stages and questions on these and connect them with straight line segments. If the students are using a computer, free flow chart software is available (see e.g. www.bestflowchart.com/flowchartsoft-ware.htm).

2. Students are given the statements and questions on pre-printed shapes and have to place them in the correct order.

3. Students are given a flow chart with mistakes in it and have to make the appropriate corrections.

4. Students are given a flow chart to follow and write up the process into a methodology.

How it works

This sequence assumes you are using the first option above.

1. Students should first identify both the starting conditions and the outcome they want to achieve by the end of the process. It is important to know when the process is complete so that they can stop the activity.

2. They should then identify all the key stages in the process and arrange them as a sequence of actions. These stages should be written in the rectangular boxes.

3. Students then create several checkpoint questions which have yes/no or true/false answers. Each decision to be made should be placed in a rhombus.

4. Each answer should lead students forwards to the next stage or back to one of the previous key stages in order that the conditions are met for moving on. For example, if a resource is missing they need to go back to the collecting stage before moving on (e.g. an essay is not complete until the references have been added).

5. Once everything is ready, they can move forward to the next step in the procedure. When the end point is reached the process stops.

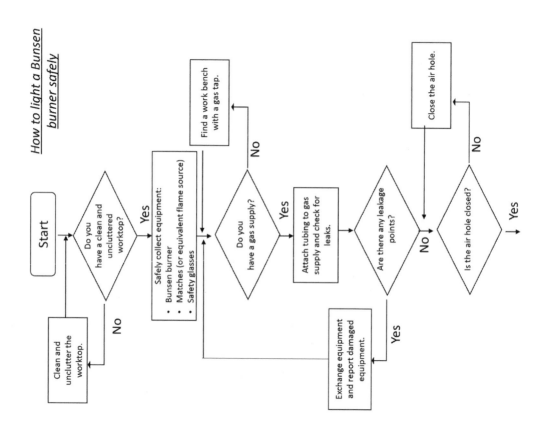

How to light a Bunsen burner safely

Start

Do you have a clean and uncluttered worktop?

No → Clean and unclutter the worktop.

Yes →

Safely collect equipment:
- Bunsen burner
- Matches (or equivalent flame source)
- Safety glasses

Do you have a gas supply?

No → Find a work bench with a gas tap.

Yes →

Attach tubing to gas supply and check for leaks.

Are there any leakage points?

Yes → Exchange equipment and report damaged equipment.

No →

Is the air hole closed?

No → Close the air hole.

Yes →

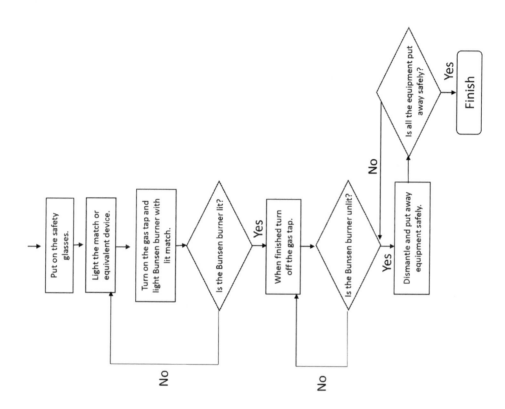

Activity types

Designing a process to make a physical object or planning an event such as a party/exhibition (emphasis on the *actions* to be taken).

Methodologies for making a significant decision, such as how to choose the best energy supplier or how to decide on a holiday that everyone in the family will be happy with (emphasis on the *decisions* to be made).

When completing a practical or common activity in class, a flow chart could be given to those students who need it to help prompt them into the next steps. However, it is important that students don't become reliant on these prompts over time.

Example activities

Science – When completing a practical task, students have to consider the logic of the methodology and take into account the possible outcomes they will be looking for at certain stages (e.g. recording the temperature of a liquid's boiling point).

English – Designing a procedure to follow when answering examination questions or when comparing texts.

Food technology – When putting together a set of instructions for a recipe.

Computing – When designing a computer program or app that will produce a clear outcome for the user.

Mathematical tips

It is important when writing out any process which leads to a definite conclusion that the information is displayed clearly and logically. This is especially pertinent with mathematical or logical reasoning and problem solving – for example, the stages in performing a complex calculation or the steps in proving a geometry theorem.

Basically, the rule is to start at the top and work downwards with each step on a new line. Alternatively, if there is room, start on the left and work across to the right. It's unlikely that anyone will get it right first time – there are always missed steps and options that only become apparent as you work through the flow chart. Therefore, get students to design the chart in rough first, check it out as they work through the process and make any amendments necessary. Alternatively, they could get another student to follow the steps to see whether it works or not, and use their feedback to update the flow chart.

Classroom Management

27. Measure of Success

Measure of Success is a rewards-based system, with teachers able to reward students differently for positive behaviours. For example, if a student picks up a pencil that another student has dropped then this kindness is worth a reward – but only at low level. At the other extreme, if someone has passed out in the corridor and a student puts them into the recovery position, sends for help and keeps calm, then that is clearly worth a much larger reward.

For students with behavioural issues this can often make a difference to their future behaviour, as it allows them to build up towards rewards over time instead of having to make one grand gesture to register on the school's reward scheme.

This exercise is designed to help students associate mathematics with positive outcomes. It should be used to reward students' positive behaviours and attitudes towards learning and their class work. It does this by rewarding them with mathematical quantities that they will need to be able to understand in everyday life, such as simple fractions, time and measures. The greater the accomplishment, the bigger the reward given to the student. Students are able to successfully build on their rewards from previous lessons and are recognised for all their achievements over time.

Preparation and resources

You will need to provide the class with miniature versions of the reward slips (see pages 162–163) to stick in their books. These can be the jug, the stars or the clock sheets.

Choose the appropriate resource according to the ability level of the class. As some students might struggle with mathematical content such as fractions, it may be better to start with the star split into quarters. Once they have gained enough rewards to fully shade in the shape, you could move them on to the more prestigious star which is divided into sixths. In this way you will be increasing the mathematical difficulty within a positive context.

You will need to decide what the reward is for fully shading in a shape. This could be a postcard home, achievement points, a prize or anything else you can think of. It is sometimes nice to do a weighted system where every shape fully shaded is equivalent to an achievement point. However, when the student has 10 shapes fully shaded it might be appropriate to send an achievement certificate or postcard home.

How it works

1. Students should stick the miniature copies of the shapes into the front of their books, which will allow them to visually build on their successes over the term and the year. This reinforces positive behaviours from all students and encourages them to give higher order answers.

2. Rewards are given in the form of the students being able to shade in their success tokens. These rewards are weighted according to the teacher's judgement. For example, if a student participates in class activity when asked a question, you could reward them by asking them to shade in a quarter of their star or five minutes of time on their clock success token. However, if a student comes to the board and presents their work and findings to the class then this deserves a greater reward, such as shading in a half of their star or 10 minutes of time on their clock success token.

3. To extend this further, when a student is allowed to shade in a segment of their shape they should note down why, so when they take their books home they can share this with their parents, guardians or carers.

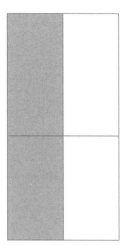
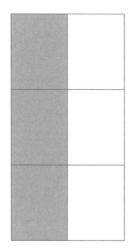
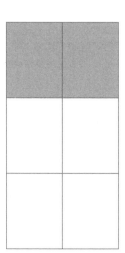

$$\frac{2}{4} = \frac{1}{2}$$ $$\frac{3}{6} = \frac{1}{2}$$ $$\frac{2}{6} = \frac{1}{3}$$

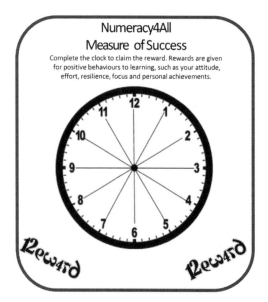

Numeracy4All
Measure of Success
Complete the clock to claim the reward. Rewards are given for positive behaviours to learning, such as your attitude, effort, resilience, focus and personal achievements.

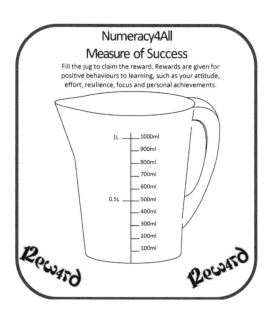

Numeracy4All
Measure of Success
Fill the jug to claim the reward. Rewards are given for positive behaviours to learning, such as your attitude, effort, resilience, focus and personal achievements.

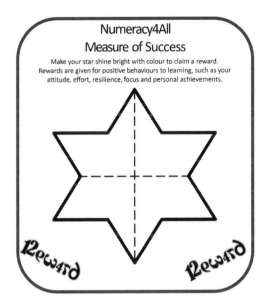

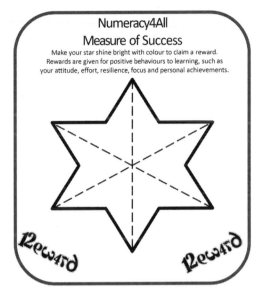

Activity types

Behaviour management

Questioning

Homework

Quality of work

Example activities

Computing – One student may have produced a basic game that fulfils the criteria, while another student may have gone above and beyond the criteria. However, both deserve some kind of reward which should be based on the amount of effort they put in and the quality of the finished product.

Mathematical tips

For the jug: there are 1,000 millilitres in 1 litre, so 0.5 litres is equal to 500 millilitres.

Equivalent fractions are when fractions represent the same quantity using different values top and bottom. For example, 50/100 = ½.

Variations

To extend the activity and make it more complex for students, deliberately use either fractional equivalents to help build their mathematical understanding (e.g. ½ instead of ²⁄₄) or measurement conversions (e.g. shade in 0.2l = 200ml of the student's jug).

28. Timers

Given a crowded curriculum and the limited length of the school term and school day, it is vital to use time effectively. During a lesson it is extremely important for students to have the opportunity to practise working to a time frame, as this is a skill they will need both for exams and in everyday life. Unfortunately, many students struggle to tell the time which can have a knock-on effect on their time management skills. Therefore, the more we use time within the classroom environment, the more the students will have a need to learn it and practise the skill.

Preparation and resources

You will need to draw a blank clock face on the whiteboard or project one using the interactive display. Some teachers find it useful to have an A3 version printed and laminated on their walls and then write on it using a whiteboard pen.

You will also need a working analogue clock on your wall so the students can calculate how much time they have left as and when necessary.

How it works

The basic principle of Timers is to put a time limit on an activity. Start with a strict time limit and see how well this works. After a trial period, ask the students whether this needs to be adjusted, and if so why.

Once the students are familiar with strict time limits, you can introduce extensions and adjustments which are based on their learned experience of doing work. There will also be occasions when time extensions are needed for contingencies that have arisen, such as when a class is struggling with a key concept or the amount of work involved was not accurately assessed beforehand. Then the students will need to add the extension on to the task finishing time to create a new finish time. Using these time frames is a good way to boost students' skills in telling the time.

1. Instead of telling the students that they have five minutes to complete an activity, tell them what the actual finish time is according to the clock on the wall – for example, 10.15 a.m.

2. Ask students to calculate how much time they have to complete the activity.

3. If a class extension is required, pause the students and either provide them with the finish time and ask them the length of time you are extending the activity by or tell students the length of the extension and ask them to calculate the new finish time.

4. By adhering to either of these two steps, you are making it clear to students that when you give a time limit, you will stick within that time frame and end at a set time (instead of running over or just forgetting about it).

Timers can also be used when a student is struggling:

1. Give students a small prompt and set a timer for one minute. The student will look again at the problem with the advice given and see if they can help themselves.

2. Check if the student has completed/made progress with the task when the time is up. If they haven't, provide more structured support and guidance. Ask the student to estimate how much time they will need, then have them record the actual time so they can judge the accuracy of their time assessment.

This promotes independence within a classroom and gives you a chance to help more students.

Activity types

Practice assessment – it would be useful for students to get used to reading a clock as they are working as they will be required to do this in an exam hall.

Exam question practice – Advise students that instead of plunging straight in, they need to look at the exam questions and think about their answers before they start writing. They should allow a set chunk of time for doing this which is related to the question's mark weighting.

Example activities

English – When completing controlled assessments.

Design and technology – When completing practical tasks, particularly in food technology, students need to be able to follow a recipe and stick to realistic timings so they can complete their cooking within the allotted lesson time.

Science – When carrying out experiments where timing is crucial.

PE – When recording times of races.

Mathematical tips

Students need to be aware there are 60 minutes in an hour. The big hand is the minute hand and the little hand is the hour hand. The numbers on a clock represent the hours and the intervals between the numbers represent five minutes of time.

Variations

Ask students to keep a record of how long various parts of a task take – for example, a science experiment takes x minutes to set up and y minutes to clean up and tidy away afterwards. These are standard processes which happen every time. In this way, they know (and you know) how long they have to do the actual experiment.

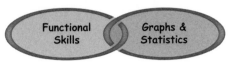
29. Weight the Task

Tasks often have several different levels of difficulty and are weighted accordingly, especially in an exam. Weight the Task is an exercise that you can use to train students from an early age to be able to spot these levels of difficulty and to allocate their time appropriately. This activity has similarities to previous activities about assessing exam questions (see Resource 2 – Writing Weigh-In) and about best use of time (see Resource 28 – Timers).

Preparation and resources

You will need a set of examination questions. These should include the marks allocated according to difficulty. Students will also need to have a clock available to look at in the classroom as they work their way through the questions.

How it works

1. When students are practising answering examination questions, ask them to monitor how long they are spending on each question (or part of a question) by recording the clock time in the margin of the page.

2. Once you have collected some data on timings from the class, ask them to calculate mean or median timings, and then hold a discussion with them on whether they think these are reasonable given the time they are allowed during an exam. How could they best use the available time to answer the questions?

3. The next time the students are attempting to answer examination questions of a similar type, you could call out the average timings so that each student can judge whether they are going too fast or too slow.

By using the actual time, instead of 'Five minutes gone' or 'Half an hour to go' statements, you will help students to secure a better understanding of clock time as well as improving their ability to pace themselves during an exam.

It is important that this activity is applied sensitively to the students in front of you. It can work with both mixed ability and streamed classes as long as there is a wide enough range of questions available.

Activity types

Exam practice

Example activities

English – Break up the reading and the writing elements of an examination task when appropriate.

Mathematical tips

Refer to the functional skills support mat (Figure 4.8) for help when working with time.

Some classes (or students who are poor at reading a clock) might benefit from having blank 12-hour clocks printed for them so they can draw on the hour and minute hands. This will help to improve their time-telling skills.

30. Groupers

Groupers provides a quick way at the start of a lesson to get students into mixed-ability groups. This activity gives students the opportunity to move around the classroom and talk to each other about what they have learned in maths to help them answer the quick questions. It also provides opportunities for them to practise mental arithmetic and problem-solving skills.

Preparation and resources

You will need to prepare and print out a question card for each student. These should be handed out before the activity. The Grouper cards can include content ranging from money and percentages to basic addition and problem-solving skills. The answers should be prepared for display on the whiteboard to avoid any confusion.

How it works

1. The aim of Groupers is for the students to calculate the value of the card based on the information given, and then to find people with the same value card as them.

2. Once the students are grouped together, you may wish to split a group of six into two threes, for example.

You can use maths questions of varying difficulty according to the ability level of students in your class. You may wish to consult with the maths department and ask them to write some quick questions based on the topics the students have recently studied or key topics the department are trying to ensure that every student understands.

Numeracy4All – Grouper

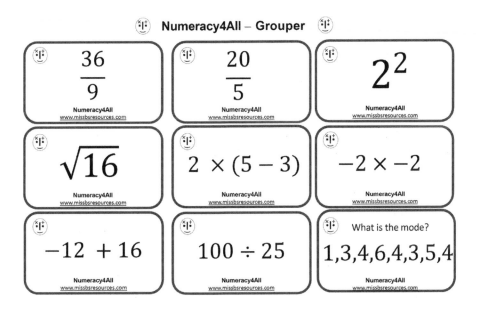

$$\frac{36}{9}$$

Numeracy4All
www.missbsresources.com

$$\frac{20}{5}$$

Numeracy4All
www.missbsresources.com

$$2^2$$

Numeracy4All
www.missbsresources.com

$$\sqrt{16}$$

Numeracy4All
www.missbsresources.com

$$2 \times (5 - 3)$$

Numeracy4All
www.missbsresources.com

$$-2 \times -2$$

Numeracy4All
www.missbsresources.com

$$-12 + 16$$

Numeracy4All
www.missbsresources.com

$$100 \div 25$$

Numeracy4All
www.missbsresources.com

What is the mode?

1,3,4,6,4,3,5,4

Numeracy4All
www.missbsresources.com

Another idea is to theme the Groupers cards around a theme or key cultural events, such as Christmas, Ramadan, Easter and Valentine's Day.

Numeracy4All – Christmas Grouper

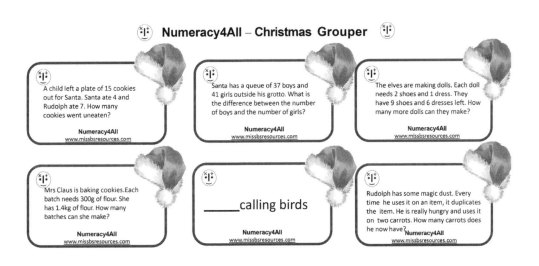

A child left a plate of 15 cookies out for Santa. Santa ate 4 and Rudolph ate 7. How many cookies went uneaten?

Numeracy4All
www.missbsresources.com

Santa has a queue of 37 boys and 41 girls outside his grotto. What is the difference between the number of boys and the number of girls?

Numeracy4All
www.missbsresources.com

The elves are making dolls. Each doll needs 2 shoes and 1 dress. They have 9 shoes and 6 dresses left. How many more dolls can they make?

Numeracy4All
www.missbsresources.com

Mrs Claus is baking cookies. Each batch needs 300g of flour. She has 1.4kg of flour. How many batches can she make?

Numeracy4All
www.missbsresources.com

_____calling birds

Numeracy4All
www.missbsresources.com

Rudolph has some magic dust. Every time he uses it on an item, it duplicates the item. He is really hungry and uses it on two carrots. How many carrots does he now have?

Numeracy4All
www.missbsresources.com

Activity types

Group activities

Example activities

Business studies – When running a group activity about a particular country, you could ask students to get into groups using that country's monetary denominations.

Drama/dance – When asking students to get into groups to put together a performance, to avoid friends ending up in the same group.

Mathematical tips

Refer to the number support mats (see Figures 4.12 and 4.13).

31. Mix Up

Sometimes, when working with groups, you want to vary who works together. It's easy to say, 'Move to the next table' when completing a carousel activity or 'Pass your sheet to the person next to you.' However, Mix Up allows you to completely mix things up without changing the activity, and without the students pre-selecting the person they will exchange their work with. And because they don't know in advance who will be looking at their work, they are likely to put more effort into their presentation to make sure it is legible – and not marked unfairly lightly by their best friend! This also helps avoid students who don't have many friends in the class from feeling left out.

Preparation and resources

No preparation is required to use this activity in its basic form. As an extension, however, you may wish to place numbers on the backs of students' chairs and then ask them to swap pieces of work or places with an odd or even numbered person. Also, for a group work activity, you can add the condition, for example, that all students who are a multiple of four should give feedback to the class.

How it works

When completing a task in which the students need to either move to another table or pass their work to another student, give a little more direction to stop them from only selecting their friends and potentially leaving certain individuals out. For example:

1. Move clockwise to the next table.

2. Share your ideas with $(2 + 1)$ or $(x + y)$ people.

3. Turn 180 degrees clockwise and pass your work to the person you are now facing. (If some pupils end up facing a wall they should pass to the person at the opposite end without anyone facing them.)

4. Move 22, n^2 or $2n$ tables anti-clockwise to your new task.

Activity types

Marking and feedback

Peer and self-assessment

Example activities

PE – When warming up and selecting partners, this is a great way to make it more random.

Drama – When asking students to peer assess each other's work, tell them they will be assessing the group sat x number of places to their left/right.

Mathematical tips

Make sure students are familiar with the terms 'clockwise' and 'anti-clockwise' and know which way to turn.

Chapter 7
Enthusiasm

Officially launching numeracy across the school is an exciting and simple way to generate a buzz among students and staff. A great place to start would be to give students the ownership of, and responsibility for, a charity fundraising event. This provides healthy competition between tutor groups and year groups. The project will extend to the students having to present their profit and loss accounts in their tutor groups, and this provides them with real-life experience of the world of business.

Organising an event or project which takes a subject to a deeper level, or to a wider audience within the community, is an excellent way of promoting student independence and engagement and encouraging their creativity and innovation. They will be developing their lateral thinking by devising challenges that bring their subject to a new level of understanding, while also calling on the fundamental skills of numeracy and literacy.

Another good way to do this would be to set up competitions which have a mathematical component. For example:

- Maths problems which are stated in words.

- Crossword puzzles which consist solely of numbers.

- 1–100 challenge – The students need to make the numbers from 1 to 100 using any mathematical operations based on a limited set of numbers, such as the year 2017. You could increase the challenge by adding the limitation that each number may only be used once.

- Design a maths clock where the numbers on the dial are replaced with mathematical sums, such as the square root of the square numbers.

- Guess the smallest positive integer – Here students have to guess a whole number between 1 and infinity that nobody else chooses. The person who picks the smallest integer is the winner. (Often students think that everyone will select the numbers 1 to 10, when in fact they don't and the winning value is often within the first ten integers.)

- Breaking a code.

- Mathematical treasure hunt – Using coordinates to locate the item to be discovered.

The essential point is that you choose appropriate competitions for the students in your class, tutor group or school. This ensures that they will achieve successes with maths, and subsequently feel more confident and comfortable engaging with numeracy-based activities.

You can also help students to see mathematics in a new light by inviting a guest speaker or teacher into the school to give talks on mathematical concepts which the students can closely relate to, such as mathematics in *The Simpsons*[1] or the use of mathematics in the creation of Pixar animation movies – there are some great videos of this on the Internet. It is important to explain the importance of mathematics in a range of professions and to explore with students the number of ways that they can have a career in mathematics.

It's also useful to have some hands-on fun with recreational maths. There are many engaging speakers and innovative companies who are set up for this pre-cise purpose, as well as a number of channels on YouTube, which look at maths with a different slant, make maths more accessible with cartoons, or set and solve problems and show how to break codes.[2] Even if you think these are too advanced for your students, they will still motivate you and get you thinking enthusiastically along new lines – and this will rub off on your students.

For those students who seem completely switched off from maths, take them away for a fun-filled weekend of activities centred on numeracy. You don't nec-essarily need to flag the numeracy element because many other subjects include a mathematical component. Adventure centres are great places to run events, with archery, rock climbing, caving and the 'leap of faith'. There are always dis-tances, heights and times that can be measured and calculated.

A good theme for the weekend would be a murder mystery. Talk to the adven-ture centre in advance and arrange to plant mathematical clues at each activity.

---

1   See S. Singh, *The Simpsons and Their Mathematical Secrets* (New York: Bloomsbury, 2014).

2   See, for example: www.numberphile.com and many YouTube maths channels such as: Mathologer
    – https://www.youtube.com/channel/UC1_uAIS3r8Vu6JjXWvastJg; Matt Parker's Stand-Up Maths
    – https://www.youtube.com/user/standupmaths; Steve Mould – https://www.youtube.com/user/
    steventhebrave; and Katie Steckles – https://www.youtube.com/user/st3cks.

Get the students to work in teams over the course of the weekend to strategically work out the clues and timeline so they can solve the murder mystery. This also provides an opportunity for a collaborative approach to learning.

Students often engage with the activities without realising that they have done any maths, and it increases their confidence to at least have a go. It is important to get the students to write up the events from the weekend, along with the facts and figures they have calculated. Stipulate that they provide a context for all the figures.

However, you don't need to take the students away to enhance collaboration between departments. When teaching ratio and proportion in mathematics, for example, liaise with staff in food technology and get the students to design and scale their recipes in a maths lesson. During the following food technology lesson, the students will then be able to make the dish. This provides them with tangible links across the curriculum, and also an enticingly scrumptious memory of the maths lesson! There are many other subjects and topics where this idea could be applied – it is just about finding the time to make the connections.

Displays are another great way to help make these links clear for students. It can be a rewarding challenge, both for teachers and for faculties, to create a corridor display clearly illustrating the way maths is used in your subject. This could range from the vocabulary that students use in a particular subject area to complex calculations or timelines. Aim to surprise students by showing them something that is a natural link, but one that is slightly obscure and which they may not have thought of before – such as information on a computing display about how smooth curves are created in computer animation by using a subdivision of midpoints.[3] This will hopefully spark their inquisitive nature to find other links.

Finally, consider having a day or a week in which to specifically celebrate mathematics. I have often managed to get all the teachers involved, so that at a set time on one day in this special week, every teacher includes a numeracy activity in their lesson. This could be as simple as a two-minute assignment or it could take up the whole lesson – for instance, in English the students designed tension graphs, in history they made human timelines, in MFL Year 7 students did number sequences in the foreign language and in PE they calculated the speed of their shots in hockey. This then generates a great conversation in form time

---

3   For more on this see: T. DeRose, Pixar: The Math Behind the Movies, *Ed.TED.com* [video] (2014). Available at: http://ed.ted.com/lessons/pixar-the-math-behind-the-movies-tony-derose.

the next day when students are asked how numerate they have been so far that week.

Hopefully, at the end of the launch or numeracy week, you will be well on your way to a holistic curriculum with numeracy and literacy at the heart of it. With the links in use, the students will – over time – become more independent and the maths basics will become more fluent from continuing practice.

You are at the very start of an exciting journey, and I wish you luck in finding out how you can best include the ideas and resources in this book in your daily practice and across the school. Together, one teacher at a time, we can make a difference in many children's lives and prepare them for life in the modern world.

List of Resources

Category	No.	Resource	Category	No.	Resource
Literacy	1	Scrabblecross	Marking and Reflection	17	Thermometer of Understanding
	2	Writing Weigh-In		18	The Real Value
	3	Mathematical Language in Extended Writing		19	Maths Marking
	4	Talk Time		20	Graph It
	5	True Value		21	Results
Exploration	6	Weight of the World	Organisation and Presentation	22	Up to Date
	7	Impact Line		23	Diagram Scales
	8	Headline Figure		24	Venn Diagrams
	9	Code Breakers		25	Two-Way Tables
Engagement	10	Twisted Figures		26	Going with the Flow
	11	Bargain Words	Classroom Management	27	Measure of Success
	12	Put the Fire Out		28	Timers
	13	Fuel Fill Race		29	Weight the Task
	14	Shopping Spree		30	Groupers
	15	Battle Words		31	Mix Up
	16	Netting Questions			

References

Cepeda, N. J., Vul, E., Rohrer, D., Wixted, J. T. and Pashler, H. (2008). Spacing effects in learning: a temporal ridgeline of optimal retention, *Psychological Science*, 19(11): 1095–1102. Available at: http://uweb.cas.usf.edu/~drohrer/pdfs/Cepeda_et_al_2008PS.pdf.

DeRose, T. (2014). Pixar: The Math Behind the Movies, *Ed.TED.com* [video]. Available at: http://ed.ted.com/lessons/pixar-the-math-behind-the-movies-tony-derose.

Garner, R. (2014). Shame celebrities who boast about poor maths, says numeracy charity, *The Independent* (15th September). Available at: http://www.independent.co.uk/news/education/education-news/shame-celebrities-who-boast-about-poor-maths-says-numeracy-charity-9734152.html.

National Numeracy (2015). *Numeracy Review*. Lewes: National Numeracy. Available at: https://www.nationalnumeracy.org.uk/sites/default/files/numeracy_review_overview_v2.pdf.

Pro Bono Economics (2014). *Cost of Outcomes Associated with Low Levels of Adult Numeracy in the UK. Pro Bono Economics Report for National Numeracy.* Available at: http://www.probonoeconomics.com/sites/default/files/files/PBE%20National%20Numeracy%20costs%20report%2011Mar.pdf.

Rohrer, D. and Taylor, K. (2007). The shuffling of mathematics problems improves learning, *Instructional Science*, 35: 481–498. Available at: http://uweb.cas.usf.edu/~drohrer/pdfs/Rohrer&Taylor2007IS.pdf.

Singh, S. (1999). *The Code Book: The Secret History of Codes and Code-Breaking*. London: Fourth Estate.

Singh, S. (2014). *The Simpsons and Their Mathematical Secrets*. New York: Bloomsbury.

UK Commission for Employment and Skills (2012). *Working Futures 2010–2020. Evidence Report 41* (August). Available at: http://webarchive.nationalarchives.gov.uk/20140108090250/http://www.ukces.org.uk/assets/ukces/docs/publications/evidence-report-41-working-futures-2010-2020.pdf.